Wilder Weather

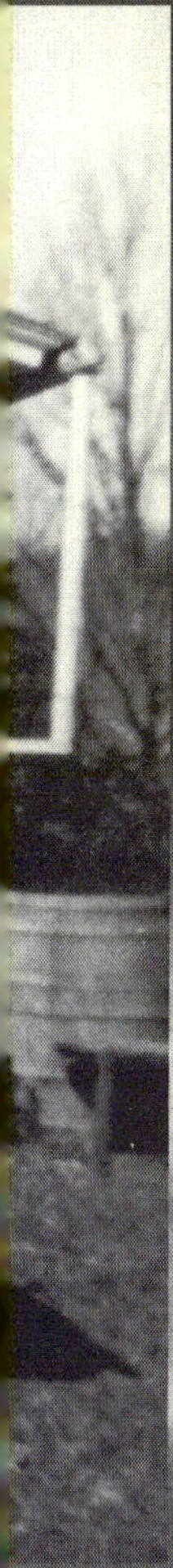

What Laura Ingalls Wilder Teaches Us About the Weather, Climate, and Protecting What We Cherish

Wilder Weather

Barbara Boustead, Ph.D.

SOUTH DAKOTA HISTORICAL SOCIETY PRESS Pierre

This publication is funded in part by Jane Page, Ruth Page Jones, Ann Page Anderson, and Mary Page.

CIP information available on request.

Printed in the United States of America

The paper in this book meets the guidelines for permanence and durability of the Committee on Production Guidelines for Book Longevity of the Council on Library Resources.

For EU customers with safety/GPSR concerns, contact:
Mare Nostrum Group, B. V., Mauritskade 21D, 1091 GC Amsterdam, The Netherlands
gpsr@mare-nostrum.co.uk

Please visit our website at sdhspress.com

Cover image: Harvey Dunn, *Storm Front*, 1933, *Courtesy South Dakota Art Museum*
Frontispiece: Laura Ingalls Wilder, age eighty-seven, framed by the doorway of her Rocky Ridge farmhouse, Mansfield, Missouri, in this 1954 photo taken by Rose Wilder Lane. *Courtesy Laura Ingalls Wilder Home Association*

Book design: Rich Hendel   Typeset in Sentinel and Miller types.

27 26 25 24 23   1 2 3 4 5 6

For Joan, Charles H., Violet,
and Charles P.—the past;

for Josh, Susan, Dawn, Samantha,
and Violet—the present;

and for Jaxon—the future.

Contents

Foreword

Try getting away from the weather in any of Laura Ingalls Wilder's classic Little House books—it can't be done. Every one of Wilder's books is permeated by meteorological forces ranging from the capricious to the catastrophic. Sun, rain, floods, drought, windstorms, hailstorms, snowstorms, tornadoes, and blizzards dominated life before forecasting, when people lived closer to the land, unable to flee from extremes of temperature and humidity. In the Little House world, the weather itself becomes, as Barb Boustead says, "a character itself—an antagonist."

Sometimes that character is a wonder and a boon, as it is when a "sugar snow" comes to the Big Woods of Wisconsin, bringing up sap in the trees and bestowing maple syrup and maple sugar cakes on the family's table. Sometimes it's terrifying, as when high winds whip prairie fires across the dry grass toward the family's Kansas cabin, driving birds, rabbits, and snakes ahead of the flames. No one who has read *The Long Winter* would argue that blizzards are not life-threatening.

As we're beginning to recognize once again, the climate controls life on Earth, a lesson Laura never forgot, from her youngest years in the Big Woods to the prairies of Kansas, Minnesota, and the Dakota Territory. No matter how hard she tried, she could not escape the long arms of fire and ice. In 1894, she and her husband Almanzo and their young daughter Rose left the drought-stricken Dakota lands after their farm failed, making the arduous horse-drawn trip to Missouri in search of a gentler clime. Instead, they spent their first winters in a windowless one-room log cabin insulated only by mud and newspapers: "Light filtered through between the logs where the chinking had fallen out," she wrote with some weariness. "So did wind and rain."[1]

Boustead is a seasoned meteorologist and climatologist with the National Oceanic and Atmospheric Administration (NOAA), and *Wilder Weather* instructs us in the science behind the Little House books' drama, showing us the difference between short-term weather conditions and long-term climate averages, between science and surmise. Anyone lucky enough to have attended one of her presentations at LauraPalooza, the Wilder conference held every two years at the townsites associated with the Ingalls and Wilder families, has been introduced to the subject's complexities, sprinkled with a little of the social magic that Wilder recalled from the town "Literaries" held when she was a teenager living in the Dakota community of De Smet. Boustead's intense enthusiasm about climate and weather—matched by a granular knowledge of Wilder's books—is infectious in the best way, and her hands-on instructional techniques have become legendary. In 2017, when LauraPalooza was held in steamy Springfield, Missouri, she had the audience make homemade ice cream in sandwich bags, a welcome treat with the hotel's air-conditioning system fighting temperatures in the nineties outside.

In *Wilder Weather*, Boustead will help readers understand the intricacies of the Fourth of July freeze of *Farmer Boy*, the unrelenting blizzards of *The Long Winter*, and the "buggy-ride tornado" of *These Happy Golden Years*. Along the way she explores all kinds of fascinating meteorological byways, using the Ingalls family and their travails as a preamble to a discussion of contemporary disasters and the essential services provided by the National Weather Service. She takes a close look at what Wilder called "grasshopper weather"—milder temperatures that Wilder (erroneously) associated with the infamous locust plague of the 1870s—and debunks folk wisdom, farmers' almanacs, weather prognosticating rodents

1. Caroline Fraser, *Prairie Fires: The American Dreams of Laura Ingalls Wilder* (New York: Metropolitan Books/Henry Holt and Company, 2017), p. 193.

(such as Punxsutawney Phil), and the woolly bear caterpillars whose fuzzy black stripes supposedly predict hard winters. She shows us that everyone can learn to be a citizen scientist by observing and recording temperatures and precipitation. She also explains how phenology works, pointing out the relationship between climatic conditions and the behavior of plants and animals that governs the seasonal patterns of everything from the first robins of spring to monarch butterflies.

Wilder Weather is an instructional marvel, but it's also wonderfully entertaining and full of surprising stories, such as the account of Wilder's adult daughter, Rose Wilder Lane, weathering the Great New England Hurricane of 1938 in Danbury, Connecticut, or Boustead's own recollection of forecasting drought as a fledgling meteorologist in farm country, or her run-ins with "fire weather" in Nebraska. Whether you are a confirmed Little House fan, a storm junkie, or simply in the market for an absorbing book about the most sensational natural spectacle on earth—the weather—you're in for a treat. "It is a beautiful world," Laura says at the end of the Little House saga, as twilight fades and the winds fall quiet, and she's right.[2]

Caroline Fraser

2. Laura Ingalls Wilder, *These Happy Golden Years*, (1943; New York: Harper & Row, 1971), p. 289.

Introduction

Everybody likes to talk about the weather!

If you've picked up this book, you probably do, too. So did Laura Ingalls Wilder. Throughout her Little House books, her columns in the *Missouri Ruralist*, and letters to daughter Rose Wilder Lane, Wilder raised the topic of the weather in casual, offhand comments and entire chapters of books. The weather is a shared interest and topic of conversation among just about any group of people.

If you ask meteorologists why they chose weather as a profession, you'll find almost all of them fall into two camps: they were intrigued by the weather they experienced when they were young, or they were afraid of the weather. I belong to the second camp. At age four, I was scared of thunderstorms, especially at night. We didn't get a lot of them in Michigan, but when we did, I would knock on my parents' door (they weren't big into bed sharing) or crawl into bed with my big sister (she was more accommodating but still got annoyed after a while). Since I already loved to read, my mother brought me to the library, hoping books about the weather would help calm my fears. I devoured every book on weather in the children's section in just months, then the weather books in the adult nonfiction section. By first grade, I announced that I wanted to be a meteorologist when I grew up. "That's wonderful!" she said, then added, "How do you even know that word?"

My parents tolerated and even encouraged my nose-pressing against the window during interesting weather (yes, they took me to the basement when it was time), my monologues on everything I was learning, and my prize-winning fourth grade science fair project on the worst tornadoes and hurricanes in U.S. history. During storms, I took my sleeping bag into the hallway because I had learned that the safest place to be during a thunderstorm is in the interior of a house, away from windows and doors.

Also during my first-grade year, Mom handed me a book she had picked up at a garage sale with two little girls peeking out the back of an old-fashioned covered wagon. I wrinkled my nose at the worn copy with a dated cover. "Just try it," Mom said with the patience of, well, Ma Ingalls herself. "I think you'll like it." She was right, as a mother usually is when she has her child's best interest at heart. I loved *Little House on the Prairie* and quickly plowed through the rest of the books in the Little House series, too.

As I grew up, I stayed close to the things I had started to love at the tender age of six. My sister and I laid out sleeping bags to watch Hurricanes Hugo (1989) and Andrew (1992) make landfall through the lens of The Weather Channel, sharing popcorn and tracking their paths on paper charts, with pencil-marked points noting latitude and longitude, pressure, and wind speed.

I kept reading the Little House books, too, like comfort food. My copy of *The Long Winter* grew warped when I took it to the beach to cool me like the lake itself. *These Happy Golden Years* warmed me with its sweet, simple love story—or charged me up about tornadoes. I found *On the Banks of Plum Creek* when I sought scientific intrigue (What were those grasshoppers, anyway?) or the adventures of an inner naughty child who was just trying her best. I went on to college, then a career in weather and climate. Over and over, I pulled the books off the shelf to read when I needed them.

I reread *The Long Winter*, for probably the hundredth time, at the tail end of the winter of 2008–2009, which held its grip in eastern Nebraska just a little too long for my liking, reminding myself that if even the Long Winter came to an end, so would mine. As during previous readings, I found myself wondering if Wilder's weather stories were anywhere near the truth.

I'm a meteorologist, I reminded myself; I know how to research past weather. And so this project began, with its many twists and turns, questions and answers and deeper questions. It blossomed into my Ph.D. dissertation, my side project, my hobby, and as of now, it is approaching two decades of work.

Many of you who pick up this book will recognize the name Laura Ingalls Wilder. Perhaps you read the Little House books as a child or watched the *Little House on the Prairie* television series, which ran on Monday nights from 1974 to 1983. Not everyone knows the real-life Wilder equally well, though. Laura Ingalls was born on 7 February 1867 to Charles ("Pa") and Caroline ("Ma") Ingalls, née Quiner, in Pepin, Wisconsin. Laura was the second of four surviving children, all girls: Mary was her older sister, Caroline ("Carrie") the next youngest, and Grace the youngest of all. Charles Frederick ("Freddie"), a son born between Carrie and Grace, died in infancy. Laura married Almanzo Wilder on 25 August 1885 in De Smet, Dakota Territory. Together, they had one daughter, Rose, on 5 December 1886, as well as a son who died in infancy. Rose had no surviving children, and none of the other Ingalls sisters had children, though Carrie was a stepmother. Thus, there are no direct descendants of the Ingalls family.

The family took to their covered wagon throughout Wilder's childhood, making six interstate moves across the Midwest and the Great Plains in her first fourteen years of life before settling in De Smet. Wilder's childhood journeys included stops in or near Independence, Kansas; Pepin, Wisconsin; Walnut Grove, Minnesota; and Burr Oak, Iowa. In adulthood, the Wilders also spent time in Spring Valley, Minnesota, and Westville, Florida, and then De Smet again, before settling for good in Mansfield, Missouri. Ma, Pa, and Mary remained in De Smet for the rest of their lives, while the remaining sisters moved away in adulthood—Grace Ingalls Dow moved just a few miles away to Manchester, while Carrie Ingalls Swanzey moved across the state to settle near Keystone.

Wilder memorialized her family's wanderlust in the Little House book series. Comprising eight books published in her lifetime and one draft manuscript published posthumously, the series spanned stories throughout her childhood and first years of marriage. Wilder first wrote the story as a memoir-style manuscript, *Pioneer Girl*, intended for an adult audience.[1] She took advice from an editor and repurposed the rejected manuscript into historical fiction books intended for children and then for young adults, embellishing and altering characters and storylines to create a better flow and narrative.[2] Her daughter, Rose Wilder Lane, was herself an accomplished writer and editor, and she took the role of copyeditor and, in some cases, collaborator. The extent of Lane's role in writing the books is robustly debated among Laura Ingalls Wilder scholars.[3]

Because the works are historical fiction based on Wilder's life, they contain a mix of fact and fictional elements created by Wilder and Lane. Scholars have dug deeply into Wilder lore to determine which elements are which, applying their expertise in subjects ranging from science and medicine to quilting and fiddling to detective work. Their deep dives into Ingalls and Wilder scholarship appear in books, documentaries, and conferences, including the academically oriented and whimsically named LauraPalooza, sponsored by the nonprofit Laura Ingalls Wilder Legacy and Research Association.[4]

Most of Wilder's deepest fictions were those of omission, as she eliminated some of the worst experiences of her childhood from the Little House narrative. For example, she omitted the birth and

1. Laura Ingalls Wilder, *Pioneer Girl: The Annotated Autobiography*, ed. Pamela Smith Hill (Pierre: South Dakota Historical Society Press, 2014).

2. The journey to publication, from the *Pioneer Girl* manuscript to the Little House series, is revealed at length in Laura Ingalls Wilder, *Pioneer Girl: The Revised Texts*, eds. Nancy Tystad Koupal and other editors of the Pioneer Girl Project (Pierre: South Dakota Historical Society Press, 2021); Laura Ingalls Wilder, *Pioneer Girl: The Path into Fiction*, ed. Nancy Tystad Koupal (Pierre: South Dakota Historical Society Press, 2023); and Pamela Smith Hill, *Laura Ingalls Wilder: A Writer's Life* (Pierre: South Dakota Historical Society Press, 2007).

3. For a few perspectives, see Hill, *Laura Ingalls Wilder: A Writer's Life*; William Holtz, *The Ghost in the Little House* (Columbia: University of Missouri Press, 1995); and John E. Miller, *Becoming Laura Ingalls Wilder* (Columbia: University of Missouri Press, 1998).

4. Laura Ingalls Wilder Legacy and Research Association (LIWLRA) LauraPalooza conference information accessed at http://www.liwlra.org/laurapalooza/.

death of her baby brother, the seedy hotel and town life of the family's year in Burr Oak, the illness that left Mary blind, and the young and unhelpful Masters family living with the Ingalls family during the Long Winter. Some of the elements in the stories, such as composite characters like Nellie Oleson, are readily broken down as fiction, the historically accurate roots identified and examined. Other fictions remain more mysterious, the edge between fact and fiction blurred by the passage of time and the loss of eyewitnesses.

Other elements are distinctly accurate in the narrative, especially for being written more than fifty years and hundreds of miles removed from most of the events and their settings. Wilder wrote to Carrie for help remembering details about De Smet in their teenage years, like Mary's dress, Sunday school hymns, and plants of the Dakota prairies, and she asked her aunt, Ma's sister Martha Carpenter, to recall everything she could about the Big Woods.[5]

The weather, however, she remembered on her own, in vivid detail. Wilder recounted with precision and accuracy the date of the first blizzard of the Long Winter, as well as the brutal winter's sudden departure and almost every detail in between. The Rocky Mountain locust infestation of the 1870s also imprinted itself deeply in her memory. As she wrote to Lane in 1938, after *On the Banks of Plum Creek* was published and while working on the manuscript of *By the Shores of Silver Lake*, "I have lived among millions of [grasshoppers]. I have seen clouds of them darken the noonday sun. I saw their bodies choke the waters of Plum Creek. I saw them destroy every green thing on the face of the earth, so far as a child would know. There are unforgettable pictures of those grasshoppers in my mind that I have tried to draw plainly in *On the Banks of Plum Creek*."[6]

The memories that affected Wilder the most were the ones she would retell with the greatest accuracy. Every book in the Little House series contains scenes centered on the weather, the climate,

5. Fraser, *Prairie Fires: The American Dreams of Laura Ingalls Wilder*, pp. 287–89, 432.

6. Laura Ingalls Wilder (LIW) to Rose Wilder Lane (RWL), 17 Aug. 1938, File 7, Box 13, Rose Wilder Lane Papers, Herbert Hoover Presidential Library, West Branch, Iowa.

or the natural environment around the family. The weather anchors scenes like the sugaring snow in *Little House in the Big Woods*, the July Fourth freeze in *Farmer Boy*, the quick-rising river in *Little House on the Prairie*, and the premise that tree claims would change the climate of the plains in *By the Shores of Silver Lake*. As the series progressed and Wilder's own attention to and memory of the weather events sharpened, weather moves from being in the scenery to becoming a character in itself—often an antagonistic one. Driven by natural and environmental disasters, the weather, the climate, and the environment take starring roles in *On the Banks of Plum Creek*, *The Long Winter*, and even in the raw manuscript-based *The First Four Years*. The fortunes of the Ingalls family ebbed and flowed with the weather—with nature—as much as they did with any other factor.

People like to share their stories of big weather. All I have to do is mention the word "thunderstorm," and whether I'm in front of a classroom of second graders or a group of small-town firefighters and EMTs, people raise their hands to share their stories. They talk about the big thunderstorm in the middle of the night that woke them up and drove them to their basements; the time they saw a funnel cloud (or maybe it was a tornado) not too far away; the time that a bunch of trees came down and they were sure it was a tornado even though it was never officially confirmed as one. The big weather story of their life might be the tornado that just missed their home or town as a child, the hurricane that flooded out a nearby neighborhood, or the snowstorm that knocked out power and made roads impassable for days and days.

When a few people gather who have experienced the same kind of extreme or catastrophic weather, the stories can spiral and build on each other, from grand tales of survival in basements and beer coolers to the subtlest details of how it smelled like freshly cut wood and greenery and natural gas in wake of the tornado. People remember how their neighbors came with chainsaws to cut a fallen tree out of their driveway so they could get out. They compare notes on how high the waters rose in their basements in a flood, how tall the snowdrifts reached against their doors and windows in a blizzard, what size the hailstones were in their yard from the big storm.

Wilder's weather experiences live up to the tallest of weather tall tales, from a six-month, blizzard-laden winter, to a three-funnel tornado, to fires, droughts, floods, bone-chilling cold, and something she called "grasshopper weather." A lifetime of practice paying attention made her weather accounts especially vivid, inviting her readers to shiver in the cold under frosty nails in the roof or wrinkle their noses with a whiff of wildfire smoke.

Trust is an intrinsic part of weather. Maybe you've wondered if you can believe Wilder's stories about the weather, or the weather observations from yesterday or a hundred years ago. Perhaps you have a favorite meteorologist, television station, app, or website you turn to for weather information. You might place your faith in an automated weather observation less—or more—than a human observation. Finding trusted sources of weather information feels like a very personal journey, one that often does not rely on statistics of accuracy or other measurable rankings.

Weather observations are measured, concrete, and tangible. Whether done by a human or a machine, the reading on the thermometer is a fact, at least within the limitations of instrument accuracy and human interpretation (sometimes that line between 55 and 56 degrees is a little fuzzy on a liquid thermometer, and it can be hard to read the difference between 0.01 and 0.02 inches of rain). The weather of the past is as clear as the instruments and humans of the past let it be to us, and their accuracy and precision have stood up well to the test of time.

As for forecasts (weather or otherwise), predicting the future is hard, even when it's based in science. Anybody who has filled out a March Madness bracket, put money in the stock market, or played a hand of poker can attest to the challenge of predicting the future. Professional weather predictions have advanced considerably through their 150 years or so of existence, but it is still an exercise in chances and probabilities, of the chaotic atmosphere, and of big differences over small spaces. Seasonal outlooks have advanced, too. They are complex and even more strongly based in chances and probabilities, making them harder for the forecasters to communicate and harder to understand and use, but they also hold value for people ranging from farmers to water resource managers.

Maybe you trust your neighbor more than anyone else about how bad a hurricane really can be where you live, and that's what informs your decision about whether to evacuate. Or perhaps you struggle to believe your grandparents about how bad the winters used to be where they live, because you can hardly even build a snowman anymore. Pulling together multiple sources can add trust, especially when their answers converge and the measurements match the memories and stories.

As much as everyone likes to talk about the weather, not as many people talk to each other about the climate. Really, climate is just the weather, but for longer. Climate is weather history. It is the weather of the past, compared to now, and the weather of the future. Climate is also the bigger patterns that nudge our daily, local weather.

Weather and climate affect us in our daily lives and the environment that surrounds us, whether that's a suburban neighborhood, a farm, the woods, an oceanfront, or a crowded city. We can see those effects if we just pay attention. You don't have to become an official weather observer (although the weather community certainly welcomes more observers!) to see the impacts in the flowers and plants, animal behaviors, and the rise and fall of waters. You can step outside and look up at the sky, catching a glimpse of clouds or distant smoke. You can peek at an outdoor thermometer or a garden rain gauge.

Just as people will talk about the big weather they've seen, they also talk about how things have changed where they live, especially if they have lived there a while. *Spring used to come later. Feels like it's been years since we had a snowy winter. Summers have sure gotten hot. Lilacs and tulips came out early this year.*

Paying attention is how we protect our families and loved ones, our neighborhoods and communities, our environment and surroundings. Paying attention to the weather keeps us tuned in when the big events are coming. Paying attention to the changes in weather over time helps us understand when things happen that have never happened before. You become one of the observers. You become part of the narrative of weather, one of the storytellers and record keepers. You learn that you can trust the weather observations because you are included in them. You empower yourself to protect what you cherish.

Wilder never wrote down how much snow fell in the October blizzard, but she did leave a trail of clues about when the blizzard occurred, and where, and among how many others. She didn't measure the rain at Plum Creek, but her stories document the hot and dry feeling as grasshoppers, crunching underfoot and devouring her family's crops, invaded her home. Her perspective has become part of the narrative others use to trace the weather and climate of her time and places. She adds color and life to the fossilized bones of weather observations.

Laura Ingalls Wilder—storyteller, children's author, farmer—becomes Laura Ingalls Wilder, weather observer.

AUTHOR'S NOTE

Throughout the book, I refer to "Wilder" or "Mrs. Wilder" when speaking of the real-life Laura Ingalls Wilder and "Laura" when speaking of her book character self. Similarly, Almanzo Wilder is "Mr. Wilder" or "Almanzo Wilder" for his real self and "Almanzo" as his fictitious self, while Rose Wilder Lane is "Lane" as her real self and "Rose" as her fictional self. Pa, Ma, Mary, Carrie, and Grace are the fictional characters; any references to the real counterparts will be as Charles Ingalls, Caroline Ingalls, Mary Ingalls, Carrie Ingalls (Swanzey), and Grace Ingalls (Dow).

I

Engage with the Records and Stories

1

Record Keepers and Storytellers

All that we know about past weather is known because record keepers and storytellers documented it.

Whether it was in a journal or letter, an official weather observation taken by a trained observer, or a graphical depiction by an American Indian storyteller, our weather has been archived into history by the pencils, paintbrushes, and keyboards of thousands of humans who included weather in their stories and routines. The same is true for climate, the average conditions of weather in the past, weather now, and weather in the future. Oral histories help fill some gaps, so long as the stories are shared. In much the same way that paleontologists can reconstruct dinosaurs that lived millions of years ago from fossilized skeletal remains, but not tell us what they sounded like, data-oriented weather observations provide scientists with information to reconstruct past weather, but the observations need fleshing out by human experience.

In other words, millions of weather observations collectively paint a picture of the weather in both the background and at the forefront of our lives. The picture will never be complete, or uniform, or sharp and clear everywhere. The observations and records are like the tiniest pieces of an enormous jigsaw puzzle, waiting to be sorted, grouped, and connected until some part of the picture begins to take shape. Even with a few gaps and missing pieces, we can see the weather (or climate) history patterns emerge, and we can see the bold strokes of big weather events when weather rises to the foreground.

Laura Ingalls Wilder demonstrated the value of the human interpretation of weather conditions through all her writings, even the simple observations she jotted down in letters to her daughter, Rose Wilder Lane. Take, for example, a letter Wilder wrote to Lane on 5 February 1937 from her farm in Mansfield, Missouri, known as Rocky Ridge. Amid an anecdote from "Manly" (her husband, Almanzo Wilder) that she thought might be useful for storytelling in her Little House books, Wilder described the weather. Under the header "Tuesday morning," which would have been 2 February—long letters were often written over the course of several days—Wilder noted, "There is snow over everything and the prospect good for more to come."[1] Though her weather account lacks the precision of an instrument weather observation, and it was not sent to the weather archives for permanent safekeeping, it does capture how the weather looked and felt, how it fit with life around it. Contrast this with the measurements from the first-order station at Springfield, Missouri, fifty miles from Rocky Ridge: "1937-02-01, Max Temperature 28, Min Temperature 12, Total Precipitation 0.12, Snowfall 1.0."[2] Decidedly less descriptive, this observation nevertheless confirms a fresh snowfall on 1 February. Wilder may have thought the prospect for more snow was good, but it materialized as just a trace of snow (visible but not enough to measure) in Spring-

1. Laura Ingalls Wilder (LIW) to Rose Wilder Lane (RWL), 5 February 1937, Box 13, Rose Wilder Lane Papers, Herbert Hoover Presidential Library (HHPL), West Branch, Iowa.

2. Unless otherwise specified, this and all raw daily and aggregated weather data used in this book are drawn from the Applied Climate Information System (ACIS), a database access system that is collectively maintained by the Regional Climate Centers (RCCs; https://www.rcc-acis.org/aboutacis_overview.html). Data in the system is collected from and quality-controlled by National Weather Service (NWS) offices around the country, then further quality-controlled and archived by the National Centers for Environmental Information (NCEI), both components of the National Oceanic and Atmospheric Administration (NOAA). For more information, see Kenneth G. Hubbard, et al., "SERVICES: A Modern

field on 3 February, with no additional measurable snow until 27 February.

Wilder's words, combined with the official weather records, bring the picture into clearer focus. Taking these observations together, we can confidently conclude that it was cold and snowy in Missouri in early February 1937.

The Backbone of Weather Records

Cooperative Observers

With the livestock already checked and the sun beginning to peer over the horizon, a farmer walks to his wide-open backyard. In the middle of the space, mounted at eye level on a metal pole, a white plastic cylinder about the size of a coffee can, with layers like a stack of upside-down plates, reflects the first pink of sunrise. The fresh snow squeaks and crunches under his boots as the farmer approaches his final chore before breakfast. He takes a metal stick with a wooden handle hanging from a nail next to the back door, labeled with tenths of an inch instead of the usual quarter-, eighth-, or sixteenth-inch graduations, and measures the fresh snowfall on a white PVC board, about eighteen inches by twenty-four inches and marked with small flags so he can find it when the snows are deep. Stooping low to see it better, he pokes the ruler through the snow several times, mentally averaging the readings, and then clears the board and sets it back between the flags. He measures again on a second board, also marked with flags, that he does not clear. Then he walks through the backyard and pokes the stick in about ten places, also mentally averaging the readings. The backyard measurements match up to his second board, adding confidence to his numbers for total snow depth. He commits the measurements of both boards to memory, then takes a metal cylinder about eight inches in diameter out of the middle of a tripod-like metal support stand, tucks it upright into the crook of his elbow, and trudges back inside.

Applied Climate Information System," *Bulletin of the American Meteorological Society* 85, no. 6 (June 2004): 811–12, DOI:10.1175/BAMS-85-6-811; and Arthur T. DaGaetano, et al., "Efficient Access to Climate Products using ACIS Web Services," *Bulletin of the American Meteorological Society* 96, no. 2 (Feb. 2015): 173–80, DOI:10.1175/BAMS-D-13-00032.1.

Shaking off his coat and gloves, he sets down the metal cylinder, then writes his snow measurements on a scrap of paper from a pile that always sits by the back door. He runs hot water in the kitchen sink, then grabs the clear plastic tube that he keeps next to the sink in winter, about two inches in diameter and also marked in tenth-inch increments. The plastic tube fits inside the metal cylinder in the warm season, when he doesn't have to melt snow, with a metal funnel that fits over the top of both the outer and inner tubes. He fills the tube with precisely 0.50 inches of hot water, then pours it into the metal cylinder. He swirls the metal cylinder until its contents turn from snow to slush to water, using his free hand to take a long pull of hot coffee. When all the ice has turned to water, he pours the liquid through the funnel and back into the smaller, clear plastic tube, setting it flat on his counter. He takes a slim metal ruler, marked in hundredths of an inch, and dips it into the plastic tube until it touches bottom. The water darkens the charcoal-gray metal to black, marking the total liquid. He writes that number down, too, subtracting the half inch he added to melt the snow. Finally, he clicks a few buttons on a digital temperature readout on the counter to view the maximum temperature, minimum temperature, and current temperature, jotting each of these on his paper, before clearing the memory on the readout to reset it for another day. He sits down at his desk next to the dining room table, coffee in hand, and logs into a website on his desktop computer. He types each reading carefully, checking twice against the numbers he wrote on his scrap of paper. He submits the report and finally can sit down to breakfast.

This farmer is a trained volunteer weather observer in the Cooperative Observer Program (COOP), one of around 8,700 volunteers who have committed to serve their country by taking weather observations at seven o'clock every morning, rain or shine, hot or cold.[3] COOP observers have kept records of temperature and pre-

3. The NWS Cooperative Observer Program (COOP) is documented extensively by the NWS (https://www.weather.gov//media/directives/010_pdfs/pd01013007curr.pdf). COOP histories and documentation of its importance through the years are available from the National Research Council, *Future of the National Weather Service Cooperative Observer Network* (Washington, D.C.:

cipitation that form the backbone of United States weather history since the organization's beginning in 1890. Many observing sites, often located on family farms, have passed down through generations. Other COOP sites are at wastewater treatment plants, state and national parks and research grounds, city suburbs, and the outskirts of small towns, where Americans live, work, play, and grow their food. Observers earn awards for their longevity of service and the quality of their measurements. They form a camaraderie with the National Weather Service (NWS) technicians who visit at least once per year to inspect and clean their gauges, make repairs, upgrade systems, and check the integrity of the observing site. The technicians make sure that grass, trees, and new buildings are not encroaching on the observing site and changing its climate without the change being recorded.[4]

This farmer's family has operated his observing site for generations. Before him, his parents and grandparents recorded their observations by pen and paper on a form, kept in a notebook and carefully tended. They would have read two glass thermometers, one filled with mercury, one with alcohol. The two thermometers were hung in a white, wooden cotton region shelter, a structure about head high and about a cubic yard in volume, with slatted sides for airflow and a door to access the instruments. While digital thermometers are preferred today, several observing sites around the country still use glass thermometers in wooden shelters. Observers read the maximum temperature from the mercury thermometer, which pauses at the highest temperature of the day until the observer spins down the mercury to reset it, flinging the thermometer

---

The National Academies Press, 1998), DOI:10.17226/6197; David Robinson, "The United States Cooperative Climate-Observing Systems: Reflections and Recommendations," *Bulletin of the American Meteorological Society* 71, no. 6 (June 1990): 826–31, DOI:10.1175/1520-0477(1990)071<0826:TUSCCO>2.0.CO;2; and Christopher A. Fiebrich, "History of Surface Weather Observations in the United States," *Earth-Science Reviews* 93, nos. 3–4 (Apr. 2009): 77–84, DOI:10.1016/j.earscirev.2009.01.001.

4. COOP site visits by NWS technicians are regulated by NWS Instruction 10-1307, last updated 20 Aug. 2019, https://www.weather.gov//media/directives/010_pdfs/pd01013007curr.pdf. The directives include guidance on how frequently to visit and system maintenance, among other topics.

like the spinner of a board game. Then, the observer uses an alcohol thermometer to read the minimum temperature, with an "index," or marker, that the alcohol pushes down to the lowest reading. The observer resets that value by moving the marker back to the current reading, also recording the temperature at observation time from the alcohol thermometer while they reset it.

Manual observation remains the most common method for daily observations. Precipitation measurements are largely unchanged from a century ago. The inner tube of the cylinder is now plastic instead of brass but uses the same calibrated ruler to measure the liquid. The metal outer tube is unchanged in modern equipment. Snowfall is still measured with a ruler marked in tenths of an inch on the best available surface, usually a white-painted wood or white polystyrene board dubbed a snow measurement board, or "snow board" for short. Observers are instructed to take ten measurements, to the nearest tenth of an inch, and average them to give a best estimate of snowfall. Observers clear the board every six hours to minimize the effects of compaction. Snow depth is also measured by ruler over a separate standard snow-measurement board. For snow depth, the snow is not cleared from the board, but rather remains until it melts away. Snow depth is measured to the nearest inch. When the snow falls in relatively still conditions, those measurements are usually fairly consistent, and the snow measurement is more reliable. In windy conditions, those measurements can vary widely, and overall confidence in having one true, representative snowfall measurement is far lower. In addition to the snowfall amount, observers collect the snow in rain gauge cylinders and melt it for a liquid equivalent of the precipitation that fell. Though there are still challenges in windy conditions—snow can be blown over the cylinder rather than falling into it—the liquid equivalent measurements are considered more reliable over time and are more commonly used for climate analysis than their frozen counterparts.

In the past, COOP observers mailed the forms every month to their NWS office, who checked the forms for accuracy and compared them to others for consistency. NWS personnel placed phone calls or wrote letters to clear up any illegible or potentially erroneous observations. Then the office mailed the collection of forms from all observers that month to the National Centers for Environmental Information (NCEI, previously known as the National Climatic Data Center). The climatologists there quality-checked the forms again, then carefully filed and archived them. Anyone seeking information about past weather observations could write or call the NCEI office and request copies, a painstaking and time-consuming process even with the most careful filing systems.

Today, the NWS office receives digital COOP observer reports nearly instantaneously via the internet. They can quality-control the day's observations right away, then look at the whole month collectively. They still transmit digital forms to NCEI at the end of each month, where the digital records are still quality-checked and carefully archived. A few observers still follow the paper process; NWS staff key the data in on behalf of the observers, then quality-control the data and transmit it to NCEI. The data records are much more accessible now, requiring only an internet connection and a bit of knowledge about which website to visit.

Sites close as observers move, die, change their land use, or become unwilling or unable to take daily readings. NWS will search for neighbors who may be willing to pick up the duties and who live close enough that the station move is negligible in the climate records. Often, though, when a site is closed, the next one is too far away to be considered a continuous record. It has its own starting date, its own history, and its own records. The move is a break in the climate records; both sites produce useful data, but a longer record is always preferred to a fragmented one.

Structural Bones

First-Order Stations

COOP observers may be the backbone of modern weather records, but they are not the only sources of data. The U.S. government keeps climate records around the country at first-order stations like the one in Springfield, Missouri—the primary weather observation sites near bigger towns and cities that define the climate records and history of the country.[5] These are sprinkled around the country, approximately evenly spaced, though with large gaps in areas of rough terrain and low population. Some of the first-order station records date back to 1870 or even earlier, while others are just a few decades old. The locations may have been relatively steady for decades or even centuries, but the keepers have changed many times, from Army personnel to government scientists to automated equipment.

U.S. government weather observations began with the establishment of the U.S. Army Signal Corps in 1870.[6] The officials assigned to weather-observing sites were almost exclusively men: civilians, often single, many relatively untrained in weather prior to their service. While some were fortunate to be assigned to more settled and established cities like Boston or Galveston, others were sent to outposts in newer, outlying locations such as Omaha or Dodge City. The observers were charged with taking observations every six hours with no backup observer providing relief.

Observers followed procedures to record temperature, precipitation, wind, snowfall, air pressure, and sky cover, along with any other weather worth documenting in notes. The notes provided insight into storms, frosts, freezes, hurricanes, meteors, fog, floods, earthquakes, and just about any other phenomena observed in the sky or on the ground. They transmitted their observations via telegraph to a central headquarters in Washington, D.C., while keeping the written records locally to pass down to the next observer as they cycled through their staff. Early meteorologists mapped and studied the observations archived in Washington, making the earliest weather analyses and predictions.

Before climatologists knew how important it was that observation sites be consistent, stations frequently moved across cities. For

5. First-order stations are defined as being staffed "in whole or part" by NWS officials and observing hourly temperature, precipitation, air pressure, humidity, wind, and sunshine, along with fixed-hour observations of clouds and weather type. Glossary of Meteorology, American Meteorological Society, https://glossary.ametsoc.org/wiki/First-order_station (2024).

6. Cleveland Abbe, "A Chronological Outline of the History of Meteorology in the United States," *Monthly Weather Review* 37, no. 4 (Apr. 1909): 146–49.

example, many official sites in cities were located on rooftops to get out of the shadows of tall buildings, until it became clear that rooftops were substandard for observation. Climatologists today know when these moves occurred and which sites were in substandard locations, and they account for these anomalies when they construct the long-term climate history of the United States. It makes tracing the moves through the years more complicated, but not impossible.[7]

In 1890, the federal government transferred its weather services to the newly created Department of Agriculture, under the title of U.S. Weather Bureau.[8] Little changed in weather operations after the transfer. Observers continued to take manual observations, telegraph them to the central headquarters, and maintain meticulous written records on-site. In 1940, the Weather Bureau moved again, this time under the aegis of the Department of Commerce.[9]

Though methods remained unchanged, the scope of weather observations shifted dramatically with the emergence of airports in the late 1930s and 1940s.[9] Weather observations became inextricably linked to commercial airports. Official observation sites moved from downtown city locations to the outskirts, where the large tracts of

7. The process of adjusting climate data to account for known station moves and instrument changes is called "homogenization" and is conducted by the National Centers for Environmental Information when preparing raw data for time series analysis (such as climate change studies). Numerous publications have addressed the process, most notably Matthew J. Menne et al., "The U.S. Historical Climatology Network Monthly Temperature Data, Version 2," *Bulletin of the American Meteorological Society* 90, no. 7 (July 2009): 993–1008, DOI:10.1175/2008BAMS2613.1.

8. National Weather Service timeline, accessed at https://www.weather.gov/timeline. The date has been documented in many places, including in the recent paper by Gregory F. Romano and others, "The Myth of Cleveland Abbe: A Review of a Manufactured History of the Creation of the National Weather Service," *Bulletin of the American Meteorological Society* 103, no. 5 (May 2022): E1390–1412, DOI:10.1175/BAMS-D-21-0022.1. This paper includes a side history of how Abbe and others may have inflated his importance in the foundation of the National Weather Service and the American Meteorological Society compared to others who also provided significant guidance and value, particularly Mr. Increase Lapham, who will arise later in this book.

9. "History of the National Weather Service," National Weather Service, https://www.weather.gov/timeline.

land needed for air travel were available. The moves disrupted the continuity between old weather records and new ones; climatologists spent decades untangling, quantifying, and smoothing the records.

In 1970, the federal weather enterprise moved again within the Department of Commerce to the newly established National Oceanic and Atmospheric Administration. This brought another name change, to the National Weather Service.[10] With the move to NWS, the workforce became more professional. Meteorologists had to meet college degree requirements. Meteorological technicians often moved into NWS after military service, where they acquired weather experience but not a college education and applied their knowledge of weather and equipment systems to both take observations and tend to the equipment. The duties of observation services in the NWS remain balanced between the positions of professional meteorologists and those of meteorological technicians, but the number of non-degreed meteorological technicians is dwindling as degreed meteorologists absorb the roles of maintaining observational systems.

As part of the NWS's Modernization Plan in the 1990s, Automated Surface Observing System (ASOS, pronounced *ay-sauce*) stations replaced paid human observers at airports across the country.[11] As early as 1979, non-first-order stations at many airports had already started to participate in the Automated Weather Observing System (AWOS, pronounced as *ay-woss*) network. ASOS stations at the first-order locations—the bigger airports—take more readings than AWOS sites, with higher-quality equipment and occasional human input. The first ASOS station came online in 1992, with all first-order sites converted by 2004. Human observers still supplement ASOS observations at 130 locations, reporting what the automated instruments sometimes cannot detect well: subtle changes in weather type, snowfall amounts, freezing rain, and weather phenomena such as funnel clouds and tornadoes.[12]

10. Fiebrich, "History of Surface Weather Observations in the United States."

11. Fiebrich, "History of Surface Weather Observations in the United States"; National Weather Service ASOS description, accessed at https://www.weather.gov/asos/asostech.

12. Scott Landolt, et al., "The Impacts of Automation on Present Weather-

Nonetheless, automated weather observations are typically more accurate than humans because the automated equipment can detect subtle changes that humans cannot. The observations provide information every few minutes, using precisely calibrated instruments that are diligently checked and maintained. Instruments can fail or provide false readings, but the error rate is no greater than with human observers. Over time, the small errors smooth out, leaving a reliable record. Temperature, dewpoint temperature (used to calculate relative humidity), air pressure, and wind speed and direction are recorded by electronic equipment that measures in terms of electrical resistance, conductivity, and interference.[13] The automated observations note the visibility and the cloud ceiling height—that is, the height of the base of the clouds above—as a service to the airports they serve. The liquid equivalent of precipitation is measured with heated tipping buckets and weighing gauges, removing the need for the human eye. Snowfall remains among the more primitive and least reliable of weather measurements, as it is still measured by a human at all observation sites.[14]

Modern weather observation systems provide consistency among observation sites across the country, ensuring that the weather is measured with the same instruments at the same calibration standards, meeting requirements for the site to be free of obstructions, and reporting at consistent times of day. Combined, COOP observers and the modern first-order stations provide a quality network of weather records: tens of thousands of pieces of evidence every day to support weather and climate stories across the country.

Type Observing Capabilities across the Conterminous United States," *Journal of Applied Meteorology and Climatology* 58, no. 12 (Dec. 2019): 2699–2715, DOI:10.1175/JAMC-D-19-0170.1.

13. Among many resources describing ASOS instrumentation, which does evolve over the years, one helpful resource is courtesy of Penn State University: https://www.e-education.psu.edu/meteo410/node/1926. NWS provides instrumentation information at https://www.weather.gov/asos/.

14. Many sources explaining the difficulties of snowfall measurement exist, including Roy Rasmussen, et al., "How Well Are We Measuring Snow: The NOAA/FAA/NCAR Winter Precipitation Test Bed," *Bulletin of the American Meteorological Society* 93, no. 6 (June 2012): 811–29, DOI:10.1175/BAMS-D-11-00052.1.

Delicate Bones

Forts and First Written Records

Predating the U.S. Army Signal Corps army outposts, sometimes called "frontier forts," took some of the first recorded weather observations in the mid- to late 1800s. Their efforts, though far less organized than Signal Corps observations, were the earliest foundations of routine, instrumental weather observations in the United States. The observers were usually the forts' doctors, the most scientific personnel available, and they kept their observations in written journals. Most outposts kept records for at least some duration while they were occupied, though there are gaps during such times as battles and engagements, seasonal abandonment, or records lost to time. Records exist from 1820 to 1892 for some 450 sites across the continental United States and Alaska. Excavated by modern climatologists, many of the old fort records have been collected and keyed into digital formats. This collection of data from the project dubbed the "19th Century Forts and Voluntary Observers Database" is conversationally called "FORTS data."[15]

If the COOP network is like the backbone of a fossilized creature, and first-order stations are the large, well-preserved bones, FORTS data are the tiny bones that are sometimes missing or hard to find—the fingers and toes, ear bones, and tails. The process to preserve them is still active, and the data are often incomplete.

Like the early COOP measurements, military observers took temperature measurements with mercury and alcohol thermometers and measured rainfall in metal cylinders with calibrated rulers. Unlike COOPs, FORTS observations often included wind speed and direction and air pressure measurements (today only automated observation stations take wind measurements). A few of the military sites even observed snow, and in rare instances, measured and reported snowfall, a rarity in nineteenth century weather obser-

15. Description of and access to FORTS data is courtesy of the Midwestern Regional Climate Center, https://mrcc.purdue.edu/FORTS. For more background on the project, *see also* Lesley-Ann Dupigny-Giroux et al., "NOAA's Climate Database Modernization Project: Rescuing, Archiving, and Digitizing History," *Bulletin of the American Meteorological Society* 88, no. 7 (July 2007): 1015–17, DOI:10.1175/BAMS-88-7-1-15.

vations. Other sites sometimes included precipitation type, which helps clarify a fuzzy picture of precipitation when temperatures hovered close to freezing.

Wind direction can be observed with a windsock, flag, or vane; nineteenth century observers estimated the direction. Wind speed, by contrast, has been observed by myriad methods through the decades. Anemometers, instruments that record wind speed, have varied through the years from mechanical spinning cups to wire sensors in a tube. Predecessors to modern anemometers required a human to watch and count the number of rotations of a device per minute to glean a wind speed, a dizzying task when the wind blew hard. Wind speeds have also been recorded by the extent and motion of a flag or tree branches in the wind and, where available, the roughness of the sea. The Beaufort scale, invented by British admiral Sir Francis Beaufort in 1804 to help sailors estimate wind speeds by land and sea indicators, is one example of estimating wind speed by sea roughness and tree movement. In FORTS records, observers often recorded winds using either something like the Beaufort scale or using a measurement that equated wind to distance, presumably the distance an object traveled on the wind over a defined period. The observations lack modern precision, but they do provide information on relative wind speeds sufficient to identify calm and windy days. Combined with wind direction and even temperature and precipitation, we can piece together this evidence to identify storms and blizzards, cold and warm fronts, wind chills, and hot and dry summer winds.

Barometers have measured air pressure through the years, another important means of piecing together weather conditions. A barometer can be homemade, fashioned out of virtually any liquid and slender tube. The aneroid barometer, invented in the 1840s, measured air pressure using a vacuum tube and expanding or contracting surfaces that moved a needle on a gauge. This likely would have been the instrument used in nineteenth century observations, including at FORTS sites. Modern automated weather observation sites house electronic pressure transducers that use highly technical electronics and polymer tubing to measure atmospheric pressure and calibrate it to the altitude at the site.

Whereas modern observers report the highest and lowest temperature in the past twenty-four hours, plus the temperature at the time of observation, FORTS observers sometimes recorded temperatures at predetermined times, such as seven in the morning and seven at night, or eight and five, or seven and three. The times varied from site to site, and even in the records of the same site. Temperatures at fixed times are unlikely to be precisely the daily high and low, though they may be close. The seven in the morning temperature is likely the daily low at most times of year, in the absence of weather that causes the temperature to drop instead of warm through the day. Daily highs are trickier, as they tend to occur closer to mid-afternoon in the winter and closer to late afternoon in the summer.

Observers took precipitation measurements at least once per day, sometimes more frequently. As in modern observations, the precipitation measurement included the melted equivalent of all precipitation that fell, whether it fell as rain, snow, ice, or a mix. The reliability of any snow measurements in the FORTS database is questionable. They do lend confidence to precipitation being in the form of snow, as well as some measure of proportion of amounts, and thus any notes or observations about snowfall are coveted, even if the precision and accuracy are low.

Adding Shape

American Indian Winter Counts and Counting Calendars

Weather observations in North America began well before European-descended settlers arrived and established their long-term weather records. Many tribes in central North America kept a pictorial calendar of their history, recorded on animal hides in spiral or row patterns and sometimes transferred later to cloth or paper.[16] Their originators referred to these as "winter counts" or "counting calendars," a means of maintaining a record of each year. Mostly begun in the 1800s, the calendar recorders may have back-

16. Documentation for this section is largely drawn from Candace S. Greene and Russell Thornton, eds., *The Year the Stars Fell: Lakota Winter Counts at the Smithsonian* (Lincoln: University of Nebraska Press, 2007), p. 1.

filled their calendars with events that had been passed along previously through oral histories.[17] The earliest Lakota count held at the Smithsonian Institution, the Battiste Good count, stretches as far back as 1700–1701.[18]

Depending on the originating tribe, winter counts and counting calendars used one or two pictures per year, highlighting the most unusual or remarkable event of the year, such as battles, killings, deaths of significant figures or significant numbers of people, disease outbreaks, hunger, animal arrivals and departures (particularly of horses), astronomical events, and weather or natural catastrophes. Only the most significant weather episodes would warrant inclusion on counting calendars, and even highly impactful weather events might be bumped by major cultural events. The absence of a weather event on the counting calendar does not indicate that it did not occur or was not important; rather, it indicates that something else was important that year, too. One event included universally across every known calendar or count was a vivid Leonid meteor shower in 1833–1834, labeled in the calendars as some variant of "the stars fell."[19] The presence of this event across all the calendars has allowed them to be dated and cataloged accurately.

The winter counts of the Oceti Sakowin—the Lakotas, Dakotas, and Nakotas—recorded events from the beginning of one winter to the beginning of the next, not just what occurred during the winter.[20] Among Lakotas, there are around a dozen preserved and archived winter counts, many of which are housed at the Smithsonian. Most calendars ceased between the late 1860s and the 1890s as the U.S. government violated treaties and eradicated American Indian customs, breaking apart tribal communities and removing or banning the tools and practices used to keep the calendars. Among the latest-running calendars, the No Ears Oglala Lakota count continued until 1919–1920.[21]

17. Ibid., p. 3.
18. Ibid., p. 13.
19. Ibid, p. viii.
20. Ibid., pp. 1–2.
21. Ibid., pp. 177–79.

Among the weather events depicted in the Oceti Sakowin calendars, many counts for 1825–1826 depict a Missouri River flood that drowned dozens of Dakota lodges.[22] The winter of 1865–1866 is noted in several calendars for many horses dying due to deep snow.[23] In contrast, 1841–1842 is in the Lakota calendars as one of three events: that deep snow fell that winter, that the Oglalas got drunk and quarreled, or that Feather in his Ear stole many horses from the Crows.[24] At least one of the Lakota counts depicted harsh winter conditions (deep snows or extreme cold) in the winters of 1720–1721 (starvation in the winter), 1722–1723 (deep snow), 1747–1748 (froze to death during the hunt), 1773–1774 (blowing snow), 1777–1778 (intensely cold winter), 1827–1828 (deep snow), and 1852–1853 (deep snow).[25] A count also noted a prairie fire in 1762–1763, from which the Brules (or Sicangus, the Burnt Thighs) took their name.[26]

The Senses and Soft Spots

Personal Accounts, Stories, and Newspapers

When verified with observations and other records, Wilder's narrative, too, becomes part of the body of evidence of weather and climate for her place and time. She is a part of the legacy of weather and climate observations. It is important to remember that Wilder wrote historical fiction, and even though it was closely based on her life experiences, it was not intended to be a historical record. For that reason, it is critical to evaluate every one of Wilder's weather and climate anecdotes, comparing it to other observations and records, before including it among the evidence for actual weather events.

For example, Wilder noted the temperature several times in her books, usually to the cold extreme. When she said it was "only twenty degrees below zero" during a Sunday sleigh ride party in *These Happy Golden Years*, was that a literal reading, or was that

22. Ibid., pp. 256–58.
23. Ibid., pp. 209–11.
24. Ibid., p. 77.
25. Ibid., p. 78.
26. Ibid., p. 84.

to distinguish it from the chilling, moody "forty below" conditions earlier that winter, as she taught school for the first time?[27] Was that "forty below" a reading or a guess? After all, the character Almanzo does mention reading the thermometer at "forty below" before he left town. If observations can corroborate one of those temperature readings, readers can gain some confidence in the accuracy of the temperatures elsewhere in the book. If the weather records verify several of the temperature readings, historians, weather researchers, and readers not only gain confidence in the temperatures, but also a written history of the temperatures in a place with only sparse weather data. They gain signposts that help mark the dates and put the events of Wilder's books into a timeline, just as they can use events from the books, like Christmas or other well-known dates, to signpost the weather events, too.

As we have seen, Wilder occasionally mentioned the weather in her letters to Lane. As both the daughter and the wife of a farmer, as well as an astute observer who had trained herself to use words to help her blind sister, Mary, visualize the world around her, Wilder was attuned to weather and its effects—and her observations were not limited to the Little House series. She kept her weather eye through adulthood.

On 25 January 1939, Wilder wrote, "We have had a rain at last. Two in. of rainfall in one day and several other rainy days besides. The creek rose up and roared! It was a warm rain soaking into the ground. Sunday night it turned cold and all night and all day and night Monday the wind blew from the southwest like the norther we met in Texas. I never before saw a wind like it here. It would freeze the marrow in ones [sic] bones. That night the wind stopped and next morning it was only 4 above zero."[28]

The official records show 3.14 inches of rain in Springfield that month.[29] Almost half the total (1.50 inches) fell on 20 January; much of the rest fell on 23–24 January (1.09 inches). But with a high temperature of 47°F and low of 40°F on 20 January, it was a stretch to call it a "warm rain." The twenty-third was warmer, with a high of

27. Laura Ingalls Wilder, *These Happy Golden Years*, p. 94.

28. LIW to RWL, 25 Jan. 1938, Box 13, RWL Papers, HHPL.

29. Via ACIS database.

53°F and a low of 42°F. The twenty-fourth had a mild maximum temperature of 52°F, but a low of 25°F and a trace of snowfall.

Weather records can help trace when Wilder wrote the letter. She mentioned that Sunday was cold. The twenty-third and thirtieth were Sundays, with the low on 23 January at 25°F. On the thirtieth, however, the low was just 9°F. The high temperature the next day was just 26°F, with a low of 3°F. These temperatures better match the "freeze the marrow in one's bones" feeling Wilder described, indicating that while she may have started writing the letter on 25 January, she added to it several days later.

Wilder's weather reports to Lane continued the next month. On 19 February she wrote, "It has rained steadily for more than a week, 3 in. in one day. The creek is roaring. Water is all over the highways and roads blocked at Ozark and Branson and Hollister. 12 ft. of water over the dam at Powersite. Up to the hotel on Rockaway beach. Snowed a little last night and looks like more rain."[30]

Wetter-than-average weather did continue into February 1938 in Springfield, with a couple tenths of an inch of rain on 6 and 10 February. From 13–18 February, however, it was especially wet. It rained on each of those days for a six-day total of 3.84 inches. The wettest day was 17 February, with 1.57 inches of rain recorded. Wilder did not note it, but six inches of snow also fell on 19 February. The combined rainfall from January to February 1938 of 7.67 inches was the thirteenth wettest on record in Springfield between 1888 and 2024.

Wilder's column, "As a Farm Woman Thinks," written for the *Missouri Ruralist*, a newspaper with statewide circulation, covered many topics related to her farm life.[31] She occasionally recalled events from her childhood, sprinkling the very early seeds that would germinate into scenes of her *Pioneer Girl* manuscript and, later, the Little House book series. Nestled in her articles about farm life, hard work, and community are observations about the weather and climate around Mansfield, changing landscapes in Wisconsin, and memories of the Hard Winter of 1880–1881 in Dakota Territory.

30. LIW to RWL, 19 Feb. 1938, Box 13, RWL Papers, HHPL.

31. These columns are reprinted in Stephen W. Hines, ed., *Laura Ingalls Wilder, Farm Journalist: Writings from the Ozarks* (Columbia: University of Missouri Press, 2008).

Wilder's words are not the only ones written about weather in the northern Great Plains and Midwest in the late 1800s. Other residents of that time and region kept journals, penned letters, wrote town histories, and contributed newspaper highlights. In the postscript of a 12 March 1937 letter, Wilder wrote to Lane, "In Pa's book is a day-by-day record of weather for Jan., Feb., March 1886. Would it be of any use to you?"[32] Pa's book is believed to be lost, perhaps purged from the Ingalls' home in De Smet after Wilder's sister Carrie died in 1941.[33] If only the book had survived, it would certainly have been of use, not just to Lane but to future weather researchers as well. Nonetheless, beginning in 1883 the *De Smet Leader/ De Smet News* issues are rife with weather notes, though again, sometimes their accuracy needs to be verified with other data to tell the complete story.

The evidence available from the wealth of historical weather data to complete the story of the Ingalls and Wilder families is scant and varies among the homesites. The Ingalls family pushed against the edge of regions settled by non-Indigenous people and often lived in locations for just a year or two at a time. The first-order stations were widely scattered and just beginning to take observations, and most of the Ingalls family homesites did not have nearby COOP observers until well into Wilder's adulthood. FORTS observations were hit-or-miss in the years they took observations, as well as in the weather parameters they observed. In some cases, the nearest observations were dozens or even hundreds of miles away. Even so, the weather records provide a skeleton of information. The combination of all the available records adds bigger-picture context and background to the weather stories in Wilder's books and life. By putting together as many of these pieces of the puzzle as possible, the picture clarifies, and the truth behind the weather stories emerges.

Wilder is a trusted storyteller, but she was not a trained weather observer, nor was she taking measurements with calibrated instruments. She instead employed words and stories to record the

32. LIW to RWL, 12 Mar. 1937, Box 13, RWL Papers, HHPL.
33. William T. Anderson, personal communication with author.

weather using evocative descriptions, or, as her sister Mary said in *Little Town on the Prairie*, she "painted pictures" with her words. Anybody can share stories of the weather and climate, and it doesn't have to be done as artfully as Wilder did to become part of the narrative. However, her artful words are the reason that so many readers of her books remember her descriptions decades after first reading them.

Citizen science remains vital to understanding and observing weather. Today, volunteer observers can purchase an inexpensive rain gauge and participate in the Community Collaborative Rain, Hail, and Snow (CoCoRaHS) Network via www.cocorahs.org, which logs precipitation measurements from thousands of people around the United States.[34] Rather than recording in paper journals that could be buried deep in boxes or lost to the ages, observers note their precipitation totals on the internet, where they are stored digitally and are accessible to the world.

Scientists and readers alike trust the account of past weather most when the stories and records converge. The contribution from each source matters, but the collective narrative is the most compelling, as it draws the account closer to truth.

34. The CoCoRaHS network is documented at its website (https://www.cocorahs.org) and in several publications, including one by Henry W. Reges, et al., "CoCoRaHS: The Evolution and Accomplishments of a Volunteer Rain Gauge Network," *Bulletin of the American Meteorological Society* 97, no. 10 (Oct. 2016): 1831–46, DOI:10.1175/BAMS-D-14-.00213.1.

2

Folklore and Forecasts

Fall is a season of change, a season of preparation and nesting. We need only take a walk on a crisp fall day to see transformation in action. Even in a suburban neighborhood, the leaves change color and drop, nuts and acorns fall to the ground for squirrels and chipmunks to collect, and the last flowers of the season burst open, then wither. Ducks and geese soar overhead, bound for warmer climes. Other animals scramble to soften and insulate their nests and store food. Insects lose flight, slow down, and devour food before they hibernate or die. In a more open place—a park, a nature preserve, or a wide expanse of rural acreage—a walk may reveal the preparation activity of larger mammals, water dwellers, and a range of native vegetation.

Live in the same area long enough and we may notice patterns through the years. The first freeze. The day the garden finally withers enough that it needs to be cut down. With a simple notebook and pen, or a digital document, anyone who pays attention can write down the dates when these backyard milestones occur, keeping track through the years.

Over time, perhaps many people living in the same area notice similar patterns in the leaves or geese or persimmon fruits. They share their stories and speculate, as people often do, about their meaning. Is it normal? Is it a sign? Is it early, or late?

Before weather models on computers, before satellites taking ocean temperatures and upper-atmosphere wind measurements, the only clues to the mystery of the weather ahead came from stories linked to the natural world. While some of this folklore has that small kernel of truth at its core, much of it is shrouded in imaginative storytelling. It is comforting and familiar, passed down through generations as local and common wisdom, something that "everybody" knows. It can be hard to challenge the common wisdom, even when evidence is clear that the folklore and the scientific observations simply do not match. After all, the rhymes and snippets are pleasing to the ears. "Ring around the moon, 'twill rain soon." "When birds fly low, expect rain and a blow." The kernels of truth make the collection of lore stickier even when some of its aspects are more far-fetched.

Signs that tell the weather ahead are particularly popular. Humans have a natural curiosity of the weather to come, born from their instinct to take shelter and stay safe from threatening weather—and, like other animals, to store food and soften the nest for the upcoming winter. While some folklore foretells the winter in the short term, just a day or two ahead, other legends foretell a season ahead.

Just as weather forecasts have brought science into the art of looking to the weather ahead for the next few days, climate forecasts (sometimes called "outlooks") have brought science into the art of looking to the collected weather for seasons ahead. Western science has been issuing climate outlooks for about fifty years and weather forecasts for about 150. Traditional Indigenous knowledge has fostered weather and climate forecasts for longer, though the information is not well connected to those outside of American Indian communities. Neither folklore nor published science, traditional ecological knowledge of weather and seasonal forecasting lies in the unexplored space between.

Weather itself is sometimes used to foretell weather to come. The first storm of a season, an early freeze, or an unusually warm or cold spell all capture our attention, sending up a warning flare that the time has come to prepare for the change of season. Maybe they remind us of how cold or stormy a winter can be, or perhaps they just make us uncomfortable, wary, and hyper-vigilant.

People often trust their own gut instincts, or the stories from their youth and from older generations. We sometimes place misguided trust in the loudest voices offering their opinions, instead of in the validated voices offering evidence-based information. We sometimes dismiss a source after one bad year when it may be the most reliable source over a collection of years.

Signs ahead of the Hard Winter of 1880–1881 abounded, at least in Wilder's tales. And it started in October 1880.

Muskrats, Geese, and Other Lore

As a farmer's daughter and a farmer's wife, Wilder no doubt heard her share of folklore for weather and seasonal forecasting. An example is the various farmers' almanacs in circulation, giving their forecasts in large chunks of time over large chunks of land, using descriptive adjectives instead of numbers and values. What does it mean to say a winter will be "cold, snowy" for a three-week period in February, especially in snowy climates? How cold must it be to qualify as "cold" in the almanac? How much snowfall qualifies as snowy? Is one snowstorm enough? What if there are a few snows but of small amounts? These questions are among the many reasons why validating farmers' almanacs is next to impossible. Much like a horoscope, people often read and interpret the intentionally vague language to fit their own experiences—a phenomenon known as confirmation bias. Moreover, no almanac publishes or publicly discusses its methodology, leaving the scientific community in the dark about its validity. Closer looks at the outcomes demonstrate that almanacs are at best as accurate as flipping a coin.[1]

1. Several studies have evaluated the farmers' almanacs' forecasts and found them unreliable or inaccurate. For example, *see* John E. Walsh and David Allen, "Testing the Farmer's Almanac," *Weatherwise* 35, no. 5 (July 2010): 212–15, DOI:10.1080/00431672.1981.9931980. *See also* Jim Gandy, "Did the Farmers Almanac Winter Forecast Verify?? Nope!," American Geophysical Union Blogosphere, Feb. 2014.

And yet, farmers' almanacs remain in circulation, still available near checkout counters at supermarkets and supply stores. Their appeal endures as humans cling to the storytelling and folklore that bring comfort, the folksy snippets that remind people of their grandparents and elders, the catchy and sometimes rhyming phrases passed through the generations.[2]

Animals, too, are part of weather lore. The groundhog—specifically Punxsutawney Phil in Gobbler's Knob, Pennsylvania—has received fame, accolades, a movie, and even a designated holiday on 2 February. According to legend, if Phil emerges from his hibernating hole and sees his shadow, he will be scared back into his den for six more weeks of winter; if he does not see his shadow, then spring is imminent. The legend is rooted in the calendar of pre-Christian Europe.[3] Like Halloween and May Day, Groundhog Day falls halfway between a solstice (winter) and an equinox (vernal, or spring). These holidays, being driven by the Earth's seasonal cycles, already had close ties to weather. A furry woodland critter foretelling the weather traces back to the eleventh century, perhaps with badgers in Phil's stead, before German immigrants looked to the more widely seen groundhogs in western Pennsylvania.

Unfortunately, climatologists have frequently debunked Phil's forecasts—in fact, his predictions are worse than chance. Following them would be less accurate than randomly rolling dice to predict an outcome.[4] Nor are they any more specific than the farmers'

2. Scott Neuman, "Decoding the Allure of the Almanac," NPR, 2 Mar. 2012, https://www.npr.org/2012/03/02/147810046/decoding-the-allure-of-the-almanac.

3. Stephen Winick, "Groundhog Day: Ancient Origins of a Modern Celebration," *Folklife Today*, American Folklife Center & Veterans History Project, Library of Congress, 1 Feb. 2022, available online at https://blogs.loc.gov/folklife/2022/02/groundhog-day-ancient-origins-of-a-modern-celebration/.

4. Several studies have evaluated Punxsutawney Phil's winter forecasts and found them unreliable or inaccurate. *See for example* National Centers for Environmental Information, "Groundhog Day Forecasts and Climate History," updated Feb. 2023, https://www.ncei.noaa.gov/news/groundhog-day-forecasts-and-climate-history; *Live Science* staff, "How accurate are Punxsutawney Phil's Groundhog Day forecasts?," *Live Science*, updated 3 Feb. 2025, https://www.livescience.com/32974-punxsutawney-phil-weather-prediction-accuracy.html.

almanacs. What does "six more weeks of winter" really mean? Is one snow in those six weeks enough to count as winter, or must the whole period average on the cold or snowy side? Is the forecast valid just for western Pennsylvania or does it extend to the rest of the eastern U.S., or to all the U.S., or all of North America? The legend is entertaining but lacks even a smidgen of truth at its core.

Though not as famous, woolly bear caterpillars are also alleged to predict the weather. Folklore suggests that the width of their black stripes or the thickness of their wool foretells the severity of the upcoming winter, and that the placement of their stripes shows when the hardest parts of winter will occur. Legend also holds that they will crawl south ahead of a hard winter or north ahead of a mild one. It turns out that their stripes have nothing to do with the weather ahead, but rather are a consequence of how long the caterpillar has been feeding, its age, and its species.[5] The thickness of their coats also is not a matter of warmth, like it is for mammals. Their fur (also known as setae) helps them freeze on purpose in hibernation to protect the core of their cells and survive the coldest of winters. And their movement is directed not by the upcoming winter's severity, but toward finding shelter under a rock or some bark, or perhaps toward finding a tasty treat.

Plants, too, join the folklore canon. Legend holds that years with lots of acorns falling (known as a "mast year") are those that bring hard winters, providing extra food for the squirrels, chipmunks, and other gatherers. It turns out that mast years are tied to the weather of the past season rather than the future one.[6] Oak trees flower in

5. A concise summary of woolly bear caterpillar characteristics by the National Weather Service in La Crosse, Wisconsin, can be found at "Woolly Bear Caterpillar: A Winter Weather Predictor or Not?," https://www.weather.gov/arx/woollybear.

6. Several studies have evaluated acorn folklore: Purdue University Extension—Forestry and Natural Resources, "Question: Why Are There So Many Acorns This Year?," *Got Nature? Blog*, 11 Oct. 2021, https://www.purdue.edu/fnr/extension/question-why-are-there-so-many-acorns-this-year/; Jay Cannon, "Are You Seeing More Acorns Falling? You Could be in the Middle of a 'Mast Year'," *USAToday*, 5 Oct. 2021, https://www.usatoday.com/story/news/nation/2021/10/05/mast-year-means-more-acorns-falling/5934916001/; Emily Moran, "Seeing a Ton of Acorns on the Ground? It Must be a 'Mast' Year for Oaks,"

the spring depending on the temperatures from the preceding winter and early spring. Mild winters without a deep spring freeze are more favorable for flowering, which translates to more acorns developing. Drought and excessive heat in the summer stymies acorn development. So, ideal weather for mast years includes a mild winter with no spring freezes, as well as a mild summer with adequate rains, along with no other major stressors in the environment. It is a waste of the tree's precious energy to sprout more seeds when the weather does not meet these conditions.

In *The Long Winter*, folklore first becomes evident by observing muskrats. Pa notes the size and thickness of the muskrat shelters around the area early in the fall. The first draft of the manuscript "The Hard Winter" accentuates the muskrat homes even more than the published book, with multiple mentions among the family members (especially Pa) of the size and thickness of their homes, repetitive almost to the point of absurdity.[7] Wilder wanted to emphasize the point that the family has paid attention to the muskrats and animals around them as a sign of the winter to come. However, if there is a relationship between muskrat house thickness and the severity of the upcoming winter, it has not yet entered the body of scientific work.

Pa notices the geese, too, flying low and fast in mid-October. Here, there is indeed a kernel of truth to his observation. Geese can fly higher in high-pressure weather, when the denser air allows them to loft more easily. On the flip side, they tend to fly lower in low-pressure weather, which typically brings more unsettled weather. Geese may migrate just ahead of a cold front, thus foreshadowing the arrival of a cold weather spell. They also might dally back north after a cold front passes. But cold weather in fall does

---

Washington Post, 23 Nov. 2019, https://www.washingtonpost.com/science/seeing-a-ton-of-acorns-on-the-ground-it-must-be-a-mast-year-for-oaks/2019/11/22/2e47f544-07ee-11ea-818c-fcc65139e8c2_story.html.

7. Laura Ingalls Wilder, "The Hard Winter" manuscript, pp. 11, 12, 18, 32, 48, 54, 148, Burton Historical Collection, Detroit Public Library. This is the original handwritten manuscript of *The Long Winter*, with a note on its cover that reads, "The Hard Winter, No. 1, Changed to The Long Winter."

not foretell the entire winter to come. While geese migration patterns can foreshadow shorter-term weather, they are not a sign of the entire winter ahead.

Not every relationship between the animal world and the winter to come has been tested, so it would be irresponsible to say that they all are mythology. Indeed, the relationships between weather or climate and the natural world (animals, plants, and ecosystems) are ripe for research to improve our understanding. It is unlikely that any one animal or plant holds the key to a winter ahead, and as we saw with the oak trees, they are more likely reacting to weather of the recent seasons or immediate future rather than the winter ahead.

Still, it is fun, and honestly, a great way to pay attention to the animals and plants in a person's own ecosystem, to look around for their cues. Anybody can watch, write it down, keep a log, and pass it along.[8] Phenologists, those who study the seasonal patterns in plants and animals, collect and track citizen observations of everything from lilac and leaf emergence to first robins to monarch butterfly sightings through websites such as the USA National Phenology Network. Each person who contributes what they have seen to the collections adds a piece to the puzzle of plant and animal behavior by seasons, which scientists can use to better understand how climate affects plants and animals.

Until researchers do their work, though, some of the relationships are yet to be understood. And they will be better understood as Western science better connects with traditional ecological knowledge to tune into the long history of observations and understanding held by Indigenous peoples.

A Blizzard in October

An early-season blizzard struck parts of South Dakota, Nebraska, Iowa, and Minnesota in mid-October 1880, reaching De Smet on the night of 14–15 October and continuing until 18 October. De-

8. USA National Phenology Network, https://www.usanpn.org/usa-national-phenology-network.

scribed widely in town histories and written journals across the region (including *The Long Winter*—"'The f-fourteenth of October, and a b-blizzard descends on us,' [Ma] said"[9]), the blizzard snowed-in trains, froze animals, and caused a few human casualties.[10] In this case, the early-season wintry weather served as a warning cry to the settlers in the region, many of whom were inexperienced with the local weather patterns (in other words, the climate.

The National Weather Service defines a blizzard as falling or blowing snow, with sustained winds or frequent gusts of 35 mph or greater, frequently reducing visibility to one-quarter mile or less, and lasting for three or more hours.[11] Wilder, of course, did not follow this modern definition of a blizzard in her books; she identified blizzards as snow driven by winds that caused visibility to drop to almost zero. It is also common in the language of the region to identify a "ground blizzard," or wind lofting snow off the ground when it is not also falling from the sky, creating near-zero visibility.[12] While the NWS does not specifically use the language "ground blizzard" in its warnings, falling snow is not a requirement for a blizzard warning, as long as the visibility is reduced by snow and wind.

October blizzards are not unheard of in South Dakota. They typically occur years or even decades apart and more commonly affect areas west of the Missouri River, which cuts through the heart of the state.[13] The October 1880 blizzard moved farther southeast than most. Weather observations around the De Smet area indicate falling temperatures and precipitation beginning on 14 October,

9. *The Long Winter* draft typescript carbon with corrections by Laura Ingalls Wilder, HHPL, p. 34.

10. Cindy Wilson, *The Beautiful Snow* (Saint Paul, Minnesota: Beaver's Pond Press, 2020), p. 39. *See also* "Winter of 1880–81, Brown County, Nebraska," https://nebraskagenealogy.com/brown/winter.htm.

11. National Weather Service, https://weather.gov/glossary/index.php?word=blizzard.

12. National Weather Service, https://www.weather.gov/safety/winter-ground-blizzard.

13. Jill S. M. Coleman and Robert M. Schwartz, "An Updated Blizzard Climatology of the Contiguous United States (1959–2014): An Examination of Spatiotemporal Trends," *Journal of Applied Meteorology and Climatology* 56, no. 1 (Jan. 2018): 173–87, DOI:10.1175/JAMC-D-15-0350.1

with temperatures falling to or below freezing on 15–17 October.[14] Precipitation during below-freezing temperatures—snow, almost certainly—continued on 16 October. With a low-pressure system centered in the four-state area of Nebraska, Iowa, Missouri, and Kansas, the snow and cold targeting South Dakota and surrounding areas were accompanied by steady winds, with a tight surface pressure gradient. Thus, any snow that fell on 15–16 October may have created ground blizzard conditions on 17–18 October, until warmer temperatures and lighter winds returned. When meteorologists at the National Weather Service anticipate the combination of snow and wind that supports blizzard conditions, they issue winter storm watches two to three days in advance of the possible blizzard. When the conditions are imminent and forecasters are more confident, they issue a blizzard warning, usually about a day in advance. Blizzards can occur in narrow bands or widespread swaths, depending on the atmospheric drivers creating the snow and wind. Snow bands can be among the trickiest forecasting puzzles, with only subtle differences from forecast to observations resulting in highly impactful differences to the people living there. Forecasters might know that a narrow band of six-inch snowfall will occur somewhere within a six-county area, but with a sharp drop-off on either side of it. Even when they are correct, folks in that area will experience a wide range of snowfall and have a wide range of perceptions about what the forecast means for them.

Even with its inherent imprecision, a forecast can provide a wealth of actionable information. Those who know a blizzard or heavy snow is coming may stock up on food or bring in extra firewood, literally or figuratively. Schools may proactively close, and businesses may send their employees home early. Highway patrols may close vulnerable interstates, require snow chains, or increase

14. FORTS observations for Fort Sisseton, Fort Bennett, Fort Randall, and Yankton, South Dakota, all report precipitation and temperatures at or below freezing on 14–16 Oct., with temperatures below freezing continuing 17–18 Oct. All four sites also reported relatively strong north-to-northwest winds 15–17 Oct. Note that wind speeds were not directly reported; instead, wind observations were included in measures of "wind movement" (reported in miles) or Beaufort wind force.

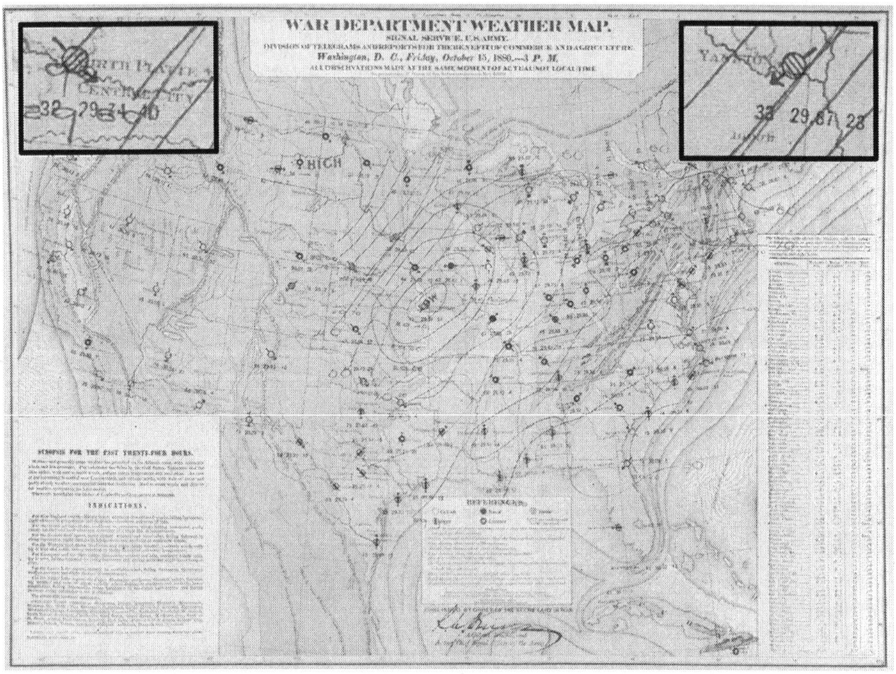

Historical weather map for 3:00 PM Eastern time on 15 October 1880. The upper left inset shows the observation for North Platte, Nebraska, with a temperature of 32°F, surface air pressure of 29.74 inches, obscured sky, and northwest winds. The upper right inset shows the observation for Yankton, South Dakota, with a temperature of 33°F, surface air pressure of 29.87 inches, obscured sky, and northeast winds. Isobars, or lines of equal pressure, show a low pressure center near the common border area of Nebraska, Kansas, Iowa, and Missouri. Courtesy National Oceanic and Atmospheric Administration

staffing, and snowplow crews set up to clear the roads. Families and communities make hundreds of cascading decisions, all based on the forecasts and warnings that meteorologists send. It is easy to complain about forecasts but also easy to remember how much harder it is to stay safe without them.

Like many settlers, the Ingalls family had no such warning and were caught unprepared by the abrupt and unusually cold and snowy October weather. After the wind tore the tar paper off the exterior of their uninsulated shanty, freezing air and snow leaked in through the cracks. To stay warm, the family marched joyfully to Pa's fiddle music, huddled under blankets, and ate warming soup.[15]

In *The Long Winter*, Pa needs little additional convincing after the blizzard and decides to move the family into their more substantial building in town for the winter, both for its thicker walls and its safer access to the town's stores and school. Ma hesitates, thinking the blizzard was perhaps a fluke, especially because the weather had turned pleasant afterwards.

Had the settlers communicated with American Indians of the region, they might have learned how often and how early the area tended to get blizzards, how long the hardest winters could last, and how much snow was usual or unusual. Indeed, Indigenous peoples around the globe have tracked winter weather patterns for generations, passing their knowledge in writing, in pictures, or orally through their own traditions. Until recent years, scientists following Western methodology (document, archive, and publish) remained largely separated from traditional ecological knowledge as by a wide moat. The patterns and outlooks noted in traditional ecological knowledge may not be widely available to read and share, but that does not erase their existence or make them invalid in the way that folklore can be debunked.

Western researchers have begun to bridge the gap and explore In-

15. Reports from Texas in Feb. 2021 include the National Centers for Environmental Information, "The Great Texas Freeze: February 11–20, 2021," https://www.ncei.noaa.gov/news/great-texas-freeze-february-2021; City of Austin and Travis County, "2021 Winter Storm Uri After-Action Review Findings Report," https://www.austintexas.gov/sites/default/files/files/HSEM/2021-Winter-Storm-Uri-AAR-Findings-Report.pdf.

digenous knowledge of the natural world more closely, introducing viewpoints not previously considered. The connections with traditional ecological knowledge are new and, in some cases, delicate, as Western scientists learn to navigate relationships with and earn the trust of Indigenous peoples. For instance, researchers have documented different types of signals used by Indigenous people to foretell winter seasons. One study that found its way into peer-reviewed scientific journals highlighted a politician in Oklahoma in 1950 and 1951 who reached out to dozens of American Indians to ask for their winter outlooks.[16] U.S. Senator Robert S. Kerr (D-Oklahoma) and his administrative assistant, Ben Dwight, a member and past principal chief of the Choctaw Nation, sent letters to thirty-three tribes across the country in 1950. These letters expressed that the senator "would like to know what some of the Indians in the various sections of the nation think about our coming winter probabilities." The letter further explained that the senator possessed a "high regard for the old Indian ways of determining such things—because they are practical and have always been able to make some very accurate predictions." Kerr and Dwight received responses with the requested predictions and some description of the basis on which those were made from the Creek Nation in Okmulgee, Oklahoma, the All Pueblo Council in Casa Blanca, New Mexico, and the Crow Agency in Lodge Grass, Montana. In 1951, the duo again sent letters, this time receiving eight responses, six of which contained winter forecasts from tribes in Arizona, Michigan, Montana, and Wisconsin. These letters did not comprise a large enough sample to test forecast accuracy, but they nonetheless provided valuable documentation of American Indian peoples who created and willingly shared winter outlooks.

Of course, none of this was in Wilder's thoughts when she included traditional ecological knowledge in her book. In *The Long Winter*, an elder Lakota man enters Harthorn's store in De Smet after the October blizzard, warning townsmen that the upcoming winter would be a very long and snowy one. Every seventh winter is

16. Randy Peppler, "'Old Indian Ways' of Predicting the Weather: Senator Robert S. Kerr and the Winter Predictions of 1950–51 and 1951–52," *Weather, Climate, and Society* 2, no. 3 (July 2010): 200–209, DOI: 10.1175/2010WCAS1055.1.

severe, he says, and every third seventh winter (that is, the twenty-first winter) is exceptionally severe and lasts especially long. The winter ahead, he cautions, is one of those third cycles of seven years. Many a reader of the Little House books counted years off by sevens to try to guess if their upcoming winter would be severe or not after reading *The Long Winter*.

Where Wilder heard the story of the Lakota elder is unclear. The story is not documented in any other known sources. He may well be an invention of the writer. In the original *Pioneer Girl* manuscript, the man is passing through town, while later versions placed him in town intentionally to warn settlers.[17] According to *Pioneer Girl: The Annotated Autobiography*, American Indians had visited white settlers in towns to talk about the weather before.[18] They often carried knowledge about history.

A similar story appears in a "literary nonfiction" publication from Rose Wilder Lane—in other words, a story based on facts but embellished for creativity, written by an author with a reputation for doing exactly that in her nonfiction work. The American Indian man in Lane's story also gives the signal for snow every seventh winter and a particularly snowy winter every twenty-first year.[19] There is also no known seven-year cycle supporting more severe winter conditions in the Dakotas or the north central United States, but the El Niño–La Niña cycle is an irregular shift of roughly three to seven years, one that will become a bone in the skeleton reconstructing the Hard Winter. If there was folklore circulated among the residents of the Dakotas around the 1880s, it is possible that a loose interpretation of the El Niño and La Niña cycles could have been the origin. Currently, there simply are not enough records and stories to compel trust in this part of Wilder's tale.

17. As noted in both *Pioneer Girl: The Annotated Autobiography* (p. 203) and *Pioneer Girl: The Revised Texts* (p. 281), the original manuscript had an "old Indian" providing warning to the townspeople of a hard winter to come while he was passing through, while later drafts had the man make an intentional trip to De Smet to warn the residents.

18. Pamela Smith Hill, *Pioneer Girl: The Annotated Autobiography*, documents the visitors in 1878 as described in the *Redwood Gazette*.

19. Nancy Tystad Koupal, "Rose Wilder Lane on Railroads and the Winter of 1880–1881," *South Dakota History* 51 no. 3 (Fall 2021): 288–313, esp. 308–9.

In any case, Wilder believed the story important enough to include in her book, and Lane in her article. Their correspondence documents how much thought went into each event included or excluded from the narrative, whether it was a record of real events or a story element to bring the book to life. Though she leaned on racial tropes in her depiction of the Lakota elder, Wilder also attributed wisdom in his words and deference to his warning by the settlers in town. The men in Wilder's story respect his traditional Indigenous knowledge and, in many cases, heed his words.

Forecasts and Modern Science

Climate forecasting and weather forecasting are about as similar as knee surgery and brain surgery—they share some anatomical roots, but the skills and specialties are quite different in application. Both types of forecasting have advanced substantially since the late 1800s, largely thanks to the power of computers and the advancement of the sciences of meteorology and climatology through research and observations.

Weather forecasts, despite the ribbing given to meteorologists, are quite accurate in places like the United States where reliable observations can feed into the computer models. Rather than waiting for a cloud in the western sky to foretell an incoming storm, meteorologists can discern days in advance with remarkable precision the locations most at risk for snow, the most likely amount of snow, and the temperatures and winds during the snow event. Most people in the United States have access to weather forecasts through television and radio stations, newspapers, websites, and phone apps, sometimes looking at multiple sources to get the range of possibilities and sometimes picking a favorite and sticking with it.

Climate forecasts (sometimes also called "outlooks") are widely available, but less frequently shared or understood than weather forecasts. Climate outlooks can show whether the atmosphere is tipping its hand toward warmer or colder conditions and wetter or drier conditions, for time windows ranging from the next week to the next month to the next season. The fundamental physics that determine much of weather also determine much of climate, but how they are applied is quite different. Rather than teasing out a day's weather forecast next month—a signal that would be drowned out by chaos and noise—climate outlooks focus on the averages over longer chunks of time. While modern forecasters still might not be able to determine before a winter occurs that it will be among the most severe to ever strike a region, they can often discern if the odds favor cold and wet conditions, and how strongly, based on the signals that science currently allows them to foresee.

The gold standard for seasonal prediction in the United States today is the National Weather Service's Climate Prediction Center. Every month, the Climate Prediction Center sends an outlook for the twelve upcoming three-month "seasons," with maps that paint areas favored for temperatures to be unusually warm or cold and precipitation to be unusually wet or dry. The outlooks are given as odds, rather than as temperatures, tilting those odds toward a favored category the same way that weighing dice can favor one roll over another. The Climate Prediction Center never forecasts a guarantee, or a 100 percent chance of anything. In the world of seasonal forecasts, there is always at least a tiny chance of all outcomes. It is too much to expect perfection.[20] Instead, the scientific process goes to work, allowing scientists to learn, adjust, and try again. The process is the reason to trust the science-based climate outlooks.[21]

The Climate Prediction Center's outlooks lean heavily on climate change trends, the influence of El Niño or La Niña, and some longer-range computer model outputs.[22] Very few global patterns have been confidently connected to outcomes for the winter ahead when looking more than a month in advance. Finding precursor signals in the atmosphere and oceans is an area of robust research, with a

20. Ed O'Lenic et al., "Developments in Operational Long-Range Climate Prediction at CPC," *Weather and Forecasting* 23, no. 3 (June 2008): 496–515, DOI:10.1175/2007WAF2007042.1

21. Annie Duke, *Thinking in Bets: Making Smarter Decisions When You Don't Have All the Facts* (New York: Portfolio/Penguin, 2019), provides insight into decisions and outcomes, namely that decisions based on a careful process can still be good decisions even when the outcome was not as hoped or expected (see for example p. 112–14).

22. O'Lenic et al., "Developments in Operational Long-Range Climate Prediction at CPC."

few promising leads but nothing that has translated into a signal reliable enough for everyday use by forecasters.

The Climate Prediction Center keeps a record of its verification, or how the forecasts compare to the outcome.[23] Climate outlooks are more skillful in some years than others, especially those "forecasts of opportunity" when signals like El Niño or La Niña are strong and provide a more reliable nudge to weather patterns, and are also more skillful in some parts of the country.[24] The Climate Prediction Center makes its forecasts, the inputs, and the outcomes all transparent, adding credibility of their methods and results.

Sometimes, the current science does not give us the tools to predict global weather patterns of winds and air pressures that either cause weather changes or show their symptoms. They sneak in like a skunk in the shadows, giving themselves away soon enough in ways that can spoil a good forecast. Scientists can often see them and label them, but only after they show themselves. In the modern weather world, some of the patterns that pushed the Hard Winter would have been ones that scientists could see in advance. Others would have been skunks in the room.

Long Winter Records and Observers

The skeleton of understanding what happened during the Hard Winter rests on the bones of the weather data. The COOP observer program was still ten years away in 1880, though a few volunteers did report weather observations to the Signal Corps Widely spaced observers gathered data at official weather stations and military forts. Stories like Wilder's add flesh to the skeleton, revealing the color and texture, the personality and character. The picture that emerges may be missing some pieces, but through Wilder's eyes,

23. CPC forecast verification is available and updated monthly at https://www.cpc.ncep.noaa.gov/products/predictions/long_range/tools/briefing/mon_veri.grid.php.

24. Robert E. Livezey and Marina M. Timofeyeva, "The First Decade of Long-Lead U.S. Seasonal Forecasts," *Bulletin of the American Meteorological Society* 89, no. 6 (June 2008): 843–54, DOI:10.1175/2008BAMS2488.1; Peitao Peng et al., "An Analysis of CPC's Operational 0.5-Month Lead Seasonal Outlooks," *Weather and Forecasting* 27, no.4 (Aug. 2012): 898–917, DOI:10.1175/WAF-D-11-00143.1.

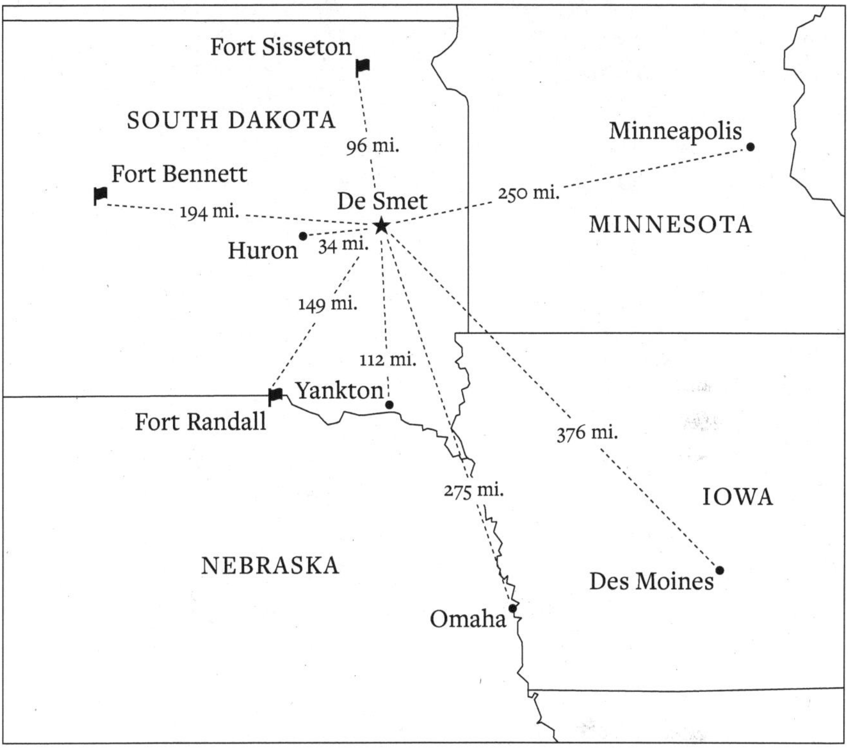

Weather observation sites available for the winter of 1880–1881 surrounding De Smet, South Dakota.

the clinical facts of the Hard Winter come to breathing, chilling life.

Wilder's was not the only account of the Hard Winter. Also dubbed the "Starvation Winter," the especially cold and snowy conditions were documented in newspapers, town and train histories, diaries, journals, and letters across much of the central to eastern United States.[25] Though Wilder recounted weather and climate events extensively throughout the Little House books, *The Long Winter* stands apart. She titled the first draft of the book "The Hard Winter," a more commonly used label in the Dakotas, but her editor determined it was too distressing for a children's book.[26] Among storytellers, Wilder's descriptions of the cold, the snow, the wind, the whiteness, and the loneliness sing as one of the most vivid and evocative accounts of the Hard Winter.

In 1880, De Smet did not yet have a first-order station. Wilder's account must be verified from weather observations as close as possible to her perspective. Omaha, Nebraska (231 miles south of De Smet), Des Moines, Iowa (277 miles southeast), and Minneapolis/St. Paul, Minnesota (214 miles east), provide the nearest long-term, continuous first-order station observations. Three military "frontier" forts in eastern present-day South Dakota took observations for at least some part of the winter: Fort Randall, near present-day Pickstown (104 miles south-southwest); Fort Bennett, under present-day Lake Oahe (155 miles west-northwest), and Fort Sisseton, near present-day Sisseton (92 miles north). The records from all these sites last only a few decades, making it a challenge to put them into long-term perspective, but they do at least give a snapshot of the Hard Winter. Yankton (105 miles south) had a FORTS record for the

25. National Academies of Sciences, Engineering, and Medicine, *Emergency Alert and Warning Systems: Current Knowledge and Future Research Directions* (Washington, D.C.: The National Academies Press, 2018), pp. 19–27, DOI:10.17226/24935; W. L. Clark, "Pioneer Days in Plymouth County," *Iowa Historical Record* 9, no. 3 (July 1893), available online at http://iagenweb.org/plymouth/Towns/Remsen.html; Ole Edvart Rølvaag, *Giants in the Earth: A Saga of the Prairie* (Harper Perennial Modern Classics, 1927).

26. Hill, *Laura Ingalls Wilder: A Writer's Life*, p. 47.

Hard Winter, along with later records from a COOP site beginning in 1932, providing at least some longer-term context to the winter's observations. Scattered observers provided a few hard-to-find records in Iowa. The official observations are a baseline to compare accounts like Wilder's, newspaper stories, and other recollections of the winter that are interpretive rather than measured.

The records include various measurements of temperatures. First-order stations and some FORTS sites gave the high and low temperatures of the day, but some FORTS sites gave the temperature at specified times. Most sites have some precipitation records, but only a few FORTS sites label the precipitation as rain or snow, and none includes snowfall measurements. The FORTS sites include some wind information, such as direction and force, but not always consistently.

Weather observations today are far less mysterious than those of the late 1800s, with rules about exactly what to measure and when, using what equipment, and to what precision. The picture they create is clearer than the observations of the past. Even so, put together, there is enough information to understand the weather of the Hard Winter, and because the details converge, they take on credibility.

Heeding the Warnings

Most people respond to warnings only after they have received them from multiple sources. Pa, as we have seen, receives warnings for a severe winter from the thick muskrat homes, the October blizzard, and the Lakota elder's warning. After the October blizzard, Pa moves the family out of their flimsy shanty and into his more substantial business building in downtown De Smet. They bring their livestock, but Pa does not have the resources to move all the hay that he and Laura had gathered to feed the animals. Instead, he intends to bring hay from the homestead to town as needed. They take their meager food and fuel supplies with them, as the first-year planting from their fields had not yielded more than a few bushels of potatoes, turnips, and other garden vegetables, along with a small amount of wheat. The family would need to supplement their food stores with food shipped in via train, at least for this first winter of homesteading. Like many settlers in the Dakotas, where trees for firewood were scarce, the Ingalls depended on coal for fuel. Some homesteaders burned old bison or cow patties for heat when necessity demanded, but coal was far more commonly used. As the family settle into their new, temporary home and brace themselves for the winter, Laura and Carrie begin attending school—one of the primary reasons Pa moves the family to town is to make it safer for them to attend through the winter—and Laura makes friends among her new classmates.

Though left out of *The Long Winter*'s narrative, in real life the Ingalls family did not live alone as the Hard Winter began. George and Maggie Masters, acquaintances from Walnut Grove (site of *On the Banks of Plum Creek*), had come for an extended visit in the spring and summer of 1880. During their stay, Maggie Masters gave birth to a baby, Arthur, in May. The young family had planned to return to Walnut Grove for the winter, but the early start to the Hard Winter stymied their departure. Wilder left them out of her story partly to enhance the Ingalls family's isolation in the narrative, and partly because they strained the Ingalls family's limited food resources, squeezed them out of the already tight spaces, and contributed little to the chores and housework.

Pa was wise to move his family to safer shelter. Around mid-November, winter returned to the region in the form of light snows and freezing rain.

Snow, of course, forms as ice crystals in the clouds and stays in that form all the way to the ground. Freezing rain falls as liquid but freezes on contact with the ground and other cold surfaces. Sleet falls as liquid and freezes into hard little pellets on its way to the ground.

The snowfall increased in frequency into December, with light snowfalls on 12–13, 19–20, and 23–26 November. After around 15 November, temperatures struggled to rise above freezing across the region. Temperatures first fell below zero as early as 17 November at the more northern sites (Minneapolis/St. Paul, Fort Sisseton, and Fort Randall), 20 November at Yankton, 21 November at Des Moines, and 5 December at Omaha, all among the earliest dates in the available records for each site, and two to four weeks earlier than normal even compared to colder late-nineteenth century climate averages.[27]

It can be challenging to know which sources to trust for weather information—the beliefs passed down through generations, the latest cutting-edge science, our own observations and sense of the atmosphere, or some combination of all of these. Big storms and weather events do often have precursor signals—like changes in wind direction, air pressure, and cloudiness—that lie at the root of much folklore. Sometimes, though, the folklore is mostly mythological, based on superstition or sheer coincidence. Weather and climate forecasting models have improved drastically over the past few decades, with sometimes uncanny ability to predict big storms. Yet they, too, can sometimes miss the mark, either in the details of the event when small differences matter a lot or, less commonly, in the big picture of unfolding weather.

Pa Ingalls, like so many people, made decisions ahead of the Long Winter to keep his family safe based on the sources he trusted. Using the best available information to keep families and communities safe is as important today as it was then. When people prepare for weather to come, they often base decisions on what they or others have experienced in that location in the past. Not every winter brings eighty inches of snow, for instance, but if it has happened in the past, it raises awareness that if the conditions are right, it could happen again. Then again, as climate shifts, some of the weather of the past will never happen again, and weather like never seen before will replace it.

Sometimes, the worst happens. The elements align to bring unprecedented extreme weather to the door of our homes, neighborhoods, and communities. When that happens, relying only on what has happened in the past or on our own personal experiences could result in being unprepared for a new extreme weather event. When those new extremes happen, it is the science of forecasts, rather than folklore and myths, that will provide accurate and potentially lifesaving advance notice and preparation.

27. Via ACIS database.

3

The Hardest Winter

Heavy snow alone does not make a blizzard. Snow can fall softly, laying down an inch or two per hour, blanketing fences, mailboxes, parked cars, roofs, and patio furniture. Under its cover, the sounds of the world are muffled. Colors dampen into shades of blue and gray and white. Visibility drops below a mile. Despite the beauty, the snow can also cause hardships, impeding travel until roads are plowed and the soft cover is breached.

In a blizzard, however, the snow does not softly pillow on porches and decks or leave white frosting on the shoulders of a coat as one walks through a winter wonderland. The tiny snow grains in a blizzard wind seem to swirl and dance in all directions, falling sideways more than straight down, and yet not slanting reliably enough to identify their direction. With the low visibility, often less than a quarter mile, the effect is disorienting. The individual snowflakes reflect away light, scattering the signals coming into our eyes that help us see light sources, shapes, and objects, creating a blinding, chaotic wall of white. Winds in blizzards are fairly straight, but they can come in gusts and puffs. As the winds wrap around buildings, trees, cars, and other obstacles, they eddy and gyrate.

The wind scours the ground of snow in exposed and

open areas, piling the snowfall in the lees of fences and trees, against houses, next to bridges, sculpting the snow piles into peaks and dips like an unbaked meringue. Patches of clear ground abruptly give way to snow-covered ravines or head-high drifts near a tree line. Areas where the snow looks shallow may be snowed over as winds continue to blow and the snows settle. Drifts may scour away as the wind shifts direction. Well-known landmarks at ground level, like roads and paths and mailboxes, disappear beneath a ceaselessly shifting coating of white, leaving those who know their way home simply unable to find it.

In a blizzard the fluffy snow crystals smash against each other, becoming pulverized into jagged, icy fragments. Each one causes a tiny scratch on exposed skin. Just a few would be irritating. A blizzard of them batters the flesh until it's raw, like an army equipped with tiny needles and stones causing thousands of microscopic abrasions. It is enough to scratch eyelids and eyeballs and make them bleed and weep. Worse, the snow melts on clothing, especially if it does not insulate the body well. The wind evaporates that water, making one feel even colder, compounding the phenomenon known as wind chill. If the air temperature is 25°F, the wind chill effect from a 30 mph wind makes it feel like 8°F. At an air temperature of 15°F, the wind chill would be 5°F.[1] The air temperature, in other words, does not have to be below zero to create biting, dangerous cold in the blizzard winds. Small wonder that many people have died or suffered due to exposure after being caught outside in a blizzard, unable to find their way to shelter.

The School Blizzard

In today's communities, when harsh winter weather looms, parents and students wait for text alerts or watch television scrolls for school closings to be announced. The decisions are often based on weather forecasts before a single flake flies. Other times, the decisions are made as snow begins to stick to roads and causes travel worries. School superintendents must balance the risk that a fore-

1. Wind chill chart available from the National Weather Service's "Understanding Wind Chill" page: https://www.weather.gov/safety/cold-wind-chill-chart.

cast does not materialize and the added costs of closures with the greater risks of dangerous roads and exposure.

Schools in Wilder's childhood could and did close due to weather, as well, but without the benefit of a weather forecast to make the decision ahead of time. The notorious "Children's Blizzard" (sometimes called the "Schoolhouse Blizzard") of 12 January 1888 devastated families across Montana, Nebraska, Dakota Territory (South Dakota), Minnesota, and surrounding areas.[2] Children and their young, female teachers went to school that day enveloped in relatively mild temperatures just above freezing. Many students likely elected to go to school wearing lighter layers and left their heaviest coats and woolen undergarments at home.

At midday, temperatures plunged to around twenty below zero on the nose of strong northwesterly winds and sudden, heavy snowfall. Some teachers sent their students home—distances up to a mile—hoping they would reach safety before conditions grew too fierce. Others sheltered in place at the schools, though the buildings lacked the fuel, insulation, food, and water to last through the duration of a storm. Even so, the choice to stay was the far safer option, and most teachers and children who stayed in place survived the blizzard. Almost every story of teachers and students trying to reach home after the storm struck ended in tragedy. When the winds and snows settled, two hundred and thirty-five children, teachers, farmers, and would-be rescuers lay dead in snow drifts, farm fields, wagons, and ditches between the schools and their homes. Many more people suffered permanent damage to their extremities due to frostbite or lost them altogether.

Leading news stories brought popular attention to the plight of families affected by the Children's Blizzard, highlighting (and possibly exaggerating) the heroic escapades of teachers. One teacher, Minnie Freeman, became subject of a poem and then a song praising her for tying her students to keep them together in the storm as they made their way to shelter. With the attention from faraway cit-

2. For an in-depth description, *see* David Laskin, *The Children's Blizzard* (HarperCollins Publishers, 2004). This book serves as the source for just about all the subsequent 1888 Children's Blizzard facts in Chapter 3. For a fictionalization, *see* Melanie Benjamin, *The Children's Blizzard* (Delacorte Press, 2021).

ies in the East came aid, but soon thereafter, as with many other disasters in the central United States, the blizzard was overshadowed by a similar event in a more populated area. The Great Blizzard of 1888 struck the East Coast from New York City to New England on 11–14 March, relegating the Children's Blizzard to obscurity.

Wilder describes a similar school blizzard in *The Long Winter*. When a snowstorm catches the town unawares, the students are trapped at school and some nearly miss their homes in their efforts to return to town. The records and stories are too incomplete to say with certainty which snow event informed Wilder's novel, but it was likely one of the earlier snowstorms, perhaps in early to mid-December 1880. While no other story or record directly supports that the event happened as Wilder described it, there is also no reason to doubt that this episode or something like it affected her and the town of De Smet. Wilder provided guideposts that help identify which blizzard may have been her schoolhouse blizzard. In the novel, Laura is cranky because Ma forces her to wear her flannel underclothes despite the warmer weather of the day, much like the children caught in the 1888 blizzard felt temperatures were mild enough to dress in fewer layers.[3] From 10 to 13 December, temperatures around the region of De Smet were rather mild, reaching the upper 40s to around 50°F at the first-order and FORTS stations. Observers reported high temperatures in the upper 30s to lower 40s on 14 December, with much colder temperatures on 15 December and a twenty-four-hour precipitation accumulation that would have fallen as snow. This precipitation was much more persistent and heavier than the several snowfalls earlier that month.

As Wilder describes it, the teacher, Miss Florence Garland, chooses to lead her students home, accompanied by townsman Mr. Foster, who comes to their aid. The group takes tenuous steps from the school toward Main Street in hopes that the buildings there will help guide them.[4] Unable to see, however, they become disoriented and veer away from the safety of the businesses along Main Street. Laura herself just barely brushes against a building on the edge of

3. Laura Ingalls Wilder, *The Long Winter* (1940; New York: Harper Trophy, 1971), p. 84.

4. Wilder, *The Long Winter*, p. 86–92.

town, shouting into the wind to her teacher and classmates to help them find their way. Her efforts allow the students and adults to gather, reorient themselves, and find their homes in town. The students, being only momentarily lost, were lucky they did not have to cross open fields or country roads.

Laura suffers from the abrasions of blizzard snow on her face, wiping her eyes and finding blood on her hand as she unwraps herself at home.[5] She and her sister, Carrie, are deeply chilled but not frostbitten, safe at home to warm up and recover.

No More Trains

Modern schools may close for a few days when the worst of winter weather strikes, but it is rare for closures to last much longer. Snowplows and tow trucks can arrive anywhere from hours to days after an especially deep snowfall to clear the roads, allowing kids and staff to return to school, shoppers to return to stores, workers to return to their businesses, and lives to move forward again. Modern-day Americans don't have to worry about being blocked from incoming food and fuel for an entire winter.

During the Hard Winter of 1880, on the other hand, the repetitive storms and accompanying biting winds dumped layer upon layer of snow and ice over the railroad tracks from Minnesota into Dakota Territory. The railroad companies tried to clear the tracks, but ultimately could not keep up with the repeated blizzards and abandoned their efforts for the season, leaving hundreds of settlements stranded for much of the winter. The exact date of the last train to reach De Smet from the east before spring is not certain, but it is very likely to have been a train that passed through on or around 19 January 1881, a train that Cindy Wilson, in her book *The Beautiful Snow*, documented to have passed through Brookings, Dakota Territory (South Dakota), some forty miles to the east.[6] Another train did reach De Smet from the west on 24 February 1881, but it carried shovelers, not food or fuel for the town.

5. Wilder, *The Long Winter*, p. 94.

6. Cindy Wilson, *The Beautiful Snow: The Ingalls Family, the Railroads, and the Hard Winter of 1880–81* (St. Paul: Beaver's Pond Press, 2020), p. 103; Cindy Wilson, personal communication with author.

When the trains stopped running, De Smet and towns like it were cut off from food, fuel, and mail, and from other communities. Without the trains, travel across wide-open spaces like the Dakotas and surrounding states could only be accomplished on foot or horse. The threat of storms limited the distance and duration that most settlers were willing to travel, because being caught in the open during a blizzard all but guaranteed severe frostbite or even death. The frequency of storms may have prevented greater death tolls, as many settlers elected to stay home.

In *The Long Winter*, the schoolhouse blizzard dumps snowdrifts into the railroad cut near Tracy, Minnesota, blocking the rail line. Pa and other men from De Smet work during a lull between blizzards to clear the cut, but another blizzard blocks the line again. That later blizzard and the resulting blockade hammers the nail in the coffin of school for the rest of the winter, as without trains to bring fuel for heat, classes cannot be held. Laura ruminates on how alone they are, despite being in town. She already feels the drudgery of being confined in the house during the blizzards, studying and trying to stay warm while projecting bravery for her sisters. One day, during a clear respite, Laura, Mary, and Carrie bundle up and go outside for fresh air. Laura urges her sisters to breathe deeply: "Cold is not so cold if you are not afraid of it."[7]

Weather observers from the closest first-order and FORTS sites to De Smet (Omaha, Yankton, Fort Bennett, Fort Sisseton, and Fort Randall) recorded precipitation with temperatures below freezing around 30 November–1 December, 4–5 December, 7 December, 11 December, and 13–14 December (potentially the school blizzard). A cold snap of 2–8 December was particularly sharp, as all sites recorded daily lows well below zero, even into the -20s°F at Fort Bennett and Fort Sisseton.

The push of Arctic air into the Dakotas is a vital ingredient of the harsh winter, and not only because it brings extreme cold. Precipitation often falls near the boundary between colder and warmer air. When cold air overtakes warmer air, as it does when a cold front moves through, the cold air forces the warmer air up and over its

7. Wilder, *The Long Winter*, pp. 116–23 (chapter titled "Alone"), p. 150.

leading edge. The moisture in the warmer air condenses into clouds, then into precipitation.

By mid- to late December 1880, at least one location in the region reported precipitation (almost certainly snow, as earlier in the month) every day from 15 to 30 December. All sites remained below freezing. In *The Long Winter*, the blizzards give way to quiet weather two or three days before Christmas, allowing the Ingalls family time to prepare. Despite the blocked railroads, mail does arrive over land, bringing a letter from Ingalls family friend and supporter Reverend Alden, promising a food barrel along with bundles of newspapers and magazines. The papers arrive with the letter, but the food barrel shipped via train, and it would not arrive for Christmas.[8] With the trains blocked, Pa, Ma, and Laura realize that finding gifts will be difficult. Ma and the girls create handmade gifts for each other. The girls put their pennies together to buy Pa a pair of suspenders, and Pa purchases Christmas candy before the stores run out of supplies. Ma gathers some of the remaining food treats—a can of oysters to make a thin, watery soup, for example—and the family celebrates with all the warmth and cheer they can muster.[9]

While the family enjoys their Christmas dinner, the next blizzard arrives.[10] Weather records indicate that snow overtook the region on 25–26 December. Northwesterly winds brought another surge of Arctic air as temperatures again plunged below zero and stayed there through 30 December. Low temperatures fell to the teens and twenties below zero, while highs stayed below freezing. The Fort Randall observer recorded winds of Beaufort force 6 and 7 on 28 and 29 December, respectively, equivalent to about 25 to 38 mph. The wind chill on those days was at best -30°F and at worst -55.

After Christmas, the Ingalls family soon runs out of coal.[11] Trees, as previously mentioned, were scarcer on the Dakota prairie in 1880 than they are today—it is not merely symbolic that Wilder used "the Lone Cottonwood" as a landmark on the prairie in her books—so wood was not available as a substitute fuel. With endless grasslands

8. Wilder, *The Long Winter*, pp. 168–73.
9. Wilder, *The Long Winter*, pp. 176–81.
10. Wilder, *The Long Winter*, p. 184.
11. Wilder, *The Long Winter*, p. 185.

around the settlers, though, there was an abundant supply of cut grass stacked as hay, and many settlers burned it, twisting loose hay by hand into sticks that would burn longer.[12] The hay sticks were a poor substitute, however, burning too quickly and giving off little heat. Twisting hay was also extremely rough on hands and clothing and had to be done almost constantly.

Without trains to replenish supplies, food stores also depleted. New farmers like the Ingalls family had harvested thin crops in 1880, and for many those food stores ran out by midwinter. Hungry settlers turned to raw wheat intended to serve as animal feed or seed for the spring planting. New towns like De Smet, lacking a flour mill, could not process the raw wheat, leaving families to mill it in hand-operated coffee grinders, monotonous, taxing, laborious work.[13]

Entertainment and diversions remain scant as the family slips into survival mode. The magazines and newspapers from the Christmas mail bundle provide fuel for their imaginations, if not for their homes or bellies. Family members take turns reading articles and story segments aloud, allowing the stories to carry longer into the winter. The exercise provides a lesson in—and appreciation for—delayed gratification. Like with their food and fuel, the family knows they have to make what they have on hand for relief and escape stretch for a long time. Another train wouldn't pass through De Smet until 4 or 5 May, and it contained no food or fuel provisions, only farm equipment and telegraph poles, just as Wilder described.[14] Another train, with a cargo of provisions, wouldn't arrive until at least a couple of days later.

The Seed Wheat Trip

With trains no longer expected until spring, and the food situation in the Ingalls' and others' households becoming dire, Wilder makes

12. For example, *see* "Hay Burner Once Kept Settlers Warm," *Farm Show* 43, no. 2 (2019): 5, https://www.farmshow.com/a_article.php?aid=34965.

13. Helen Bergh, "Early History of Brown County, South Dakota" (Western Printing Co., 1970), http://files.usgwarchives.net/sd/brown/ehbc/ehbc-fulltext.txt.

14. Wilson, *The Beautiful Snow*, p. 240.

a rare storytelling shift away from Laura, into Almanzo's perspective. Almanzo makes the case to his brother, Royal, that selling his seed wheat as food will not be enough to save the town and would prevent him from being able to plant it the following spring. Almanzo proposes a trip to seek out a farmer, Anderson, who is rumored to have a healthy stockpile of wheat.[15]

At the very first LauraPalooza conference in July 2010, Jim Hicks, a retired physics teacher from Illinois, gave a presentation on the physics of Almanzo's seed wheat trip.

Speaking after Hicks, I mentioned that I, too, had looked at the seed wheat trip and tried to identify a probable date it could have occurred, based on the moon cycle and cold and clear weather in a gap between blizzards. I explained that according to FORTS data, snow fell in Omaha and Yankton on 11 February, in Yankton on 12–13 February, in Omaha, Yankton, and Fort Bennett on 14–15 February, and at Fort Randall on 15 February. The following day, 16 February, was a cold, clear day between snows at all sites, with lows in the double digits below zero and highs between -3 and 12°F across the sites. Light snow fell again on 17 February, with another clear day on 18 February and in fact several clear days before the next prolonged snowfall on 22–24 February.

Based on the evidence, then, I concurred with Hicks that 16 or 18 February were probable (but not certain) dates, with an optimized combination of weather and astronomical factors.

Hicks and I had never met, had never corresponded, had not spoken a word to each other before our presentations. Here, in front of a live audience, the very best of science took the stage: two passionate and excited scientists with the hearts of teachers, humming with the resonance of two scientific tunes in delightful harmony, using compatible but not exactly the same methods to converge on an answer to a burning question many Wilder enthusiasts had held for years.

For the rest of that conference and at subsequent LauraPaloozas for years to come, curious attendees brought questions to both of us. What else did we know? How did we find our answers? Had we

15. Wilder, *The Long Winter*, pp. 255–59, 275.

thought about investigating this other thing, or that? Hicks and I found each other to converse, as well, during and in between future LauraPaloozas. We were ecstatic about the agreement of our findings—and relieved, honestly—and equally excited about the scientific curiosity sparked in the audience. Hicks and I continue to put our heads together to answer some of the scientific riddles left by Wilder.

The date of the trip aside, Hicks calculated that a one-horse sled large enough to haul around fifteen bushels of wheat would have been able to travel up to nine miles in about six to seven hours on snow when fully loaded—that is, on the return trip after finding the wheat. The sleds would have been lighter and faster on the trip out. His calculation took into account human riders walking alongside the sleds at least some of the time to ease the burden on the horses, as well as helping the animals out of holes when they fell through weak spots in the drifted snow.[16] The distance is a far cry from the "twenty miles" each way that Wilder indicated repeatedly in her writing, but she did have a tendency to overestimate distances (she wrote that the *Little House on the Prairie* homesite was "forty miles" from Independence when it was actually a much shorter twelve miles away; her inability to physically locate the Kansas homesite forced her to estimate distance based on memory).[17]

Hicks concluded that the homestead, if it existed, was probably less than nine miles south-southeast of De Smet.[18] Niels Anderson (also documented as Nels Anderson, Niels Anderson Kopperrud, Niels A. Kopperrud, and Nelson Kopperrud), it turns out, owned a homestead about six miles south-southeast of De Smet.[19] As Wilder described the landscape of the homestead, when Almanzo and Cap arrive from the north, they descend a slope toward a flat valley and across a slough. Circling to the south of the home, they spot a door

16. Jim Hicks, personal communication with author.

17. Wilder, *The Long Winter*, pp. 244, 258, 272.

18. Jim Hicks, personal communication with author.

19. Record for Neils Anderson (6 SSE De Smet—section NW 26-110-56), U.S. Census, De Smet Township, Kingsbury County, Dakota Territory, 14 June 1880, p. 14; U.S. Census, Baker Township, Kingsbury County, South Dakota, 12 June 1900, p. 6.

in the snowbank.[20] Hicks's photographs of the homestead site, taken on a visit in 2001, also show this landscape. According to Hicks,

> To the northwest of Anderson's NW ¼ section 26:110N, Range 56 W, there is indeed a ridge that looms over the entire Anderson homestead. There is low ground leading up to the homestead from the east and northeast, which is the route Cap and Almanzo could have taken; that is, just west of Lakes Henry and Thompson. If one observes the NE corner of Anderson's NW ¼ homestead, a traveler might follow the flatter ground available to the west, but just north of the property, and travel to the homestead house by coming in from the south. The entire elevation of the Anderson homestead is low with a slight downward tilt toward the east, the Lake Thompson side.[21]

Homestead papers filed in June 1887 reveal that Anderson pre-empted (purchased) his claim on 27 May 1881, as well as the adjacent land. He built a new home on the claim and raised five acres of oats in 1881, just as the fictional Anderson suggested: "'Well,' he said finally, 'I might sow some oats.'"[22]

In *The Long Winter*, when Almanzo and Cap return to town with the wheat, Mr. Loftus initially marks up the wheat substantially, angering the hungry residents of De Smet. After a confrontation and negotiation, Loftus agrees to sell the wheat at cost. The De Smet residents ration the seed wheat based on need, allowing each person to purchase a fair share to help them make it through the winter. Though repetitive, the seed wheat provides a staple for their diets through the rest of the winter, staving off starvation. Whether the scene is true to the experience in De Smet or not, other resources indicate that sharing and distributing food by need occurred in several cut-off communities during the Hard Winter.[23]

20. Wilder, *The Long Winter*, p. 274.

21. Jim Hicks, personal communication with author.

22. *See* "Homestead Proof—Testimony of Claimant" in Homestead file for Final Certificate, No. 4996, Neils Anderson, at Mitchell, Dakota Territory, 27 May 1881; Wilder, *The Long Winter*, p. 278.

23. William Stennett, *Yesterday and To-Day: A History* (Rand McNally & Company, 1905), p. 65.

Notably, Almanzo Wilder's obituary mentions his heroic seed wheat trip, but by itself this does not prove that the trip occurred.[24] Fact and fiction have blurred more than one obituary in history. Nonetheless, the combination of stories and weather, census, and homestead claim records suggests that Almanzo Wilder's seed wheat trip was at least feasible.

The Chinook Arrives

Wilder's narrative in *The Long Winter* begins with sharp details and distinct events: the October blizzard, the school blizzard, and the Christmas evening blizzard. As winter continues and the trauma compounds, the narrative is increasingly blurred and fragmented. From around mid-January through the end of February, there are vague mentions of one storm after another, with a clear day or sometimes a few clear days in between. Indeed, records from around the region indicate multiple snow events throughout the six-week period from mid-January through February. Only three days in those six weeks were clear of falling snow at all sites in the region; otherwise, snow relentlessly raked across the observing sites, frequent, sometimes substantial, and often with wind.

Our brains work to protect us from harm and trauma. A traumatic event can cause several responses, including magnifying the trauma or dissociating to forget it, even simultaneously intensifying and fragmenting the memory.[25] With the hunger, cold, and fear that Wilder felt during the Hard Winter, it was clearly traumatic to Wilder and the Ingalls family. Moreover, chronic cold and hunger make it difficult to remember details, which is why kids in poverty struggle to learn.[26] Laura faces this same dilemma in one of their

24. Almanzo Wilder's obituary, *Mansfield Mirror*, 23 Oct. 1949.

25. "Can You Unconsciously Forget an Experience?" Science*Daily*, Texas A&M University, accessed 27 Dec. 2023, www.sciencedaily.com/releases/2016/12/161209081154.htm; "The Impact of Trauma onAdult Sexual Assault Victims, Part III: How Trauma Affects Memory and Recall," Department of Justice Canada, accessed 27 Dec. 2023, https://www.justice.gc.ca/eng/rp-pr/jr/trauma/p4.html.

26. There are many resources to support this finding, including: Eric Jensen, *Teaching with Poverty in Mind* (ASCD, 2009); H. B. Ferguson, S. Bovaird, and

at-home recitations, struggling to remember a verse she previously knew well.[27]

As winter seemed to loosen its grip in March and early April, the breaks between snows were often several days long. The high temperature even briefly poked above freezing a few times, enough to melt a little snow. Wilder did not mention these days in *The Long Winter*, possibly because she had lost sight of them through the overpowering shadow of the cold winter or because they did not fit the narrative she wanted to construct. Even with the slight foreshadowing of spring in these above-freezing days, weather records show that cold and snowy winter weather continued into mid-April across the region.

And then, a switch flipped. In the middle of a mid-April night, Laura hears the wind. But rather than bringing the scratching and clawing of snow and ice, it carries the sound of dripping water. It is a warm, melting wind, which Laura correctly identifies.

"Pa! Pa! The chinook is blowing!"[28]

Pa teases her the next morning for yelling about the wind after listening to it blow all winter long, but she retorts that it was the chinook, and that makes all the difference.

Also known as "the snow-eater wind," the chinook provides Laura with a good reason to be excited. These winds feel warmer and cause the snowpack to melt and evaporate. Chinooks are unique to flat lands east of a mountain chain, especially a north-to-south chain with a westerly wind blowing over it, such as the Rocky Mountains and the high to central Great Plains, which provide the ideal geography for chinook winds. As the winds rise on the western slope, any moisture they carry is shed. Dry air descends the eastern slope, compressing and warming in the process due to the increasing air pressure. The same change in pressure causes our ears to pop when we rapidly descend in altitude, such as when an airplane descends for landing.

M. Mueller, "The Impact of Poverty on Educational Outcomes for Children," *Pediatrics & Child Health* 12, no. 8 (Oct. 2007): 701–6, DOI:10.1093/pch/12.8.701.

27. Wilder, *The Long Winter*, p. 253.

28. Wilder, *The Long Winter*, p. 312.

This combination of warmth due to compression, drying, and the air movement caused by the wind itself is especially conducive to evaporating snow on the ground, because warmer, drier air can melt more snow and take on more atmospheric moisture. The wind also carries the moisture away so the cycle can continue.

Weather records around the region confirm the abrupt, extreme swing to warmer weather. While temperatures had risen above freezing only a few times from mid-March through mid-April, on 13–14 April the high temperature rose above freezing at all of the FORTS sites in South Dakota and for the rest of that spring never again failed to reach above freezing during the day. The last below-freezing temperatures at night at any sites in the region occurred on 21 April.

With a deep snowpack in place, on clear and calm nights, it is difficult for nighttime temperatures to stay above freezing. In a process called radiational cooling, the surface of the snow loses thermal energy, the heat of day escaping into the atmosphere above. Drier air and lower humidity would make the air especially prone to radiational cooling and thus more likely to fall below freezing. Because the nighttime temperatures did not fall below freezing from mid-April into the summer, it is unlikely that conditions were clear and calm. Instead, it was likely breezy and possibly cloudy. Temperatures would be more likely to stay warm if the winds were out of the west (the chinook), southwest, or south.

In the FORTS records, indeed, winds prevailed from the south, southeast, or southwest almost every day from 13 to 30 April and were often gusty. Computer model reconstructions of weather patterns confirm unusually strong winds from the south across Kansas, Nebraska, and South Dakota in mid- to late April.[29] The persistence

29. Reconstructions from computer models throughout this book use the 20th Century Reanalysis version 3, available online at https://psl.noaa.gov/cgi-bin/data/composites/plot20thc.day.v2.pl and documented in Laura C. Slivinski et al., "Towards a More Reliable Historical Reanalysis: Improvements for Version 3 of the Twentieth Century Reanalysis System," *Quarterly Journal of the Royal Meteorological Society* 145, no. 724 (2019), https://doi.org/10.1002/qj.3598. Hereafter, citations will reference the model reconstructions as 20thCv3.

of both day and nighttime winds aided the rapid melting of the unusually deep snowpack after the Hard Winter of 1880–1881.

Sudden melting of an especially deep snowpack means only one thing: the water had to go somewhere, and it did. Rapidly.

The Hard Winter's Spring Floods

Snowmelt flooding in the spring of 1881 earns a mere one-line mention in Wilder's novel. Pa explains to Ma that a wagon would not be able to travel over land to haul supplies before the tracks were thawed and dug out of the ice and snow, because "all the roads are under water, the sloughs are lakes in every direction, and even on the upland, a wagon would mire down in the mud."[30]

Though largely omitted from the Little House books, the spring floods in 1881 were as record-setting as the antecedent winter, and they were a direct result of it. By the time warmer weather arrived in April, the depth of the snowpack across the northern Plains and upper Midwest was measured in feet. With the abrupt transition to above-freezing temperatures, the snow began to melt suddenly and continuously. The resultant flooding, which peaked that month, affected the entire stretches of the Mississippi and Missouri Rivers, along with every major tributary in the network of waterways that extends from Montana and Wyoming to Wisconsin and Illinois and down to the Gulf of Mexico. South Dakota state historian Doane Robinson noted in 1904 that in 1881, "while the drifts and the ice remained unbroken by the spring sunshine in the Dakota region the breakup occurred at the usual season in the upper country, thus precipitating upon the lower region a winter flood."[31]

The initial melt waters drained into rivers that, given the long duration of cold temperatures, were covered with exceptionally thick ice. The rising water lifted the ice, which broke apart and began to move downstream. Ice chunks caught in the flow scoured whatever lay in their path. Jams developed where the ice became congested and stopped moving, impeding the water flow and sending it up and over or around the jam. These jams often happen quickly, with

30. Wilder, *The Long Winter*, p. 317.

31. Doane Robinson, *History of South Dakota*, vol. 1 (B.F. Bowen & Co., 1904), p. 397.

telltale signs such as water levels dropping downstream. Both in the past and in modern times, officials have resorted to explosives to clear stubborn ice jams during the worst congestion. The initial rounds of ice jam flooding in March and early April 1881, and especially on 27–30 March, erased low-lying river towns like Green Island, Nebraska.[32] The booming town of Vermillion, South Dakota, lost one-third of its buildings in the floods, causing its citizens to relocate farther up the bank when it was rebuilt.[33] The smaller town of Niobrara, Nebraska, also relocated to higher ground as a result of a series of three ice jam floods in March and April, moving building by building to a site one and a half miles away.[34]

The second round of flooding overwashed the Missouri and Mississippi River basins, raising the rivers to record flood levels that would not be surpassed for many decades. Near Omaha and Council Bluffs, the Missouri River changed its course, leaving an oxbow lake now known as Lake Manawa.[35] In southwestern Illinois, late-April flooding of the Mississippi destroyed most of the town of Kaskaskia.[36] The river also rerouted east of town, rather than west, making it one of the only places in Illinois that lies west of the Mississippi River.[37]

32. Mark Hilton, "Flood of 1881 Leaves Lost Settlements in its Wake: The Story of Green Island," Historical Marker Database, accessed 28 Dec. 2023, https://www.hmdb.org/m.asp?m=177901.

33. Arthur L. Rusch, "Portrait of a River Town: Vermillion before the Flood of 1881," *South Dakota History* 40, no. 4 (2010).

34. Shelly Kulhanek, "Niobrara River and its flooding hasn't defeated little town yet," *Lincoln Journal Star*, 20 Mar. 2019, https://journalstar.com/news/state-and-regional/nebraska/niobrara-river-and-its-flooding-hasnt-defeated-little-town-yet/article_0bbcb5b8-b3e0-516b-af58-8b87702c162d.html.

35. Kristine Gerber and Ryan Roenfeld, *Council Bluffs: 365 Days, 150 Years* (Nonpareil LLC, 2007), https://semspub.epa.gov/work/07/30022860.pdf.

36. Tim Crosby, "SIU researchers seek lost community: the original site of Kaskaskia," *Southern Illinois University News*, 15 Feb. 2023, https://news.siu.edu/2023/02/021523-SIU-researchers-seek-lost-community-the-original-site-of-Kaskaskia.php.

37. "Kaskaskia Village, Illinois," U.S. Census 2020, accessed 11 Jan. 2024, https://data.census.gov/profile?g=160XX00US1739129. The town has been plagued by repeated floods, including one in 1993, and dwindled to a population of just a couple dozen people.

The flooding also further delayed the trains from restoring service to the communities already hard-hit by the winter. Across eastern South Dakota and western Minnesota, trains either had not yet resumed service or had only made a run or two after the snow blockades melted away and cleared a path, before rivers flooded tracks, washed out bridges, and impeded the railways again.[38] Wet and muddy roadways made overland travel difficult, too, as Pa had expressed to Ma in the novel. The flooding in the spring of 1881 significantly delayed the recovery of the communities hit hard by the Hard Winter.

The Tally

From the October blizzard to the April chinook, the Hard Winter of 1880–1881 spanned six months, one of the longest on record, bringing persistent cold and recurring snowfall to much of the central United States. Various stories indicate that a total of eleven to fourteen feet of snow "on the level" fell in eastern South Dakota and surrounding areas, or between 132 and 168 inches.[39] That total, if accurate, by far exceeds any measured snowfall records in the region, where the record highest amounts are 95 inches in Sioux Falls (1968–1969), 90 inches in Huron (2000–2001), 73 inches in Brookings (1968–1969), and 71 inches in Yankton (2015–2016). The sudden end of the cold and snow in 1881 brought relief from the weather itself, but the aftereffects lasted for months. The scars lasted for generations.

After the October blizzard, the Ingalls family discover cattle that need ice broken away from their faces, as well as a bird that has buried itself in a haystack and needs a little encouragement to fly south. The cattle discovery by the family, whether symbolic or real,

38. Numerous reports of railroad bridges damaged by flooding appear in April 1881. For instance, the *Janesville (MN) Argus* on 26 Apr. 1881 reported, "Railroad trestle this side of New Ulm washed out and also bridge further west on the Cottonwood. The west is again shut off." Other reports of flood damage came from Pierre (*Brookings County Press*, 31 Mar. 1881); Egan (*Moody County Enterprise*, 21 Apr. 1881); Ortonville (*Grant County Review*, 21 Apr. 1881); and between Yankton and Sioux City (*Sioux Falls Times*, 19 Apr. 1881), just to list a few.

39. Stennett, *Yesterday and To-Day: A History*, p. 65.

highlighted one of the worst outcomes of the October blizzard: thousands of cattle and sheep across the plains perished during the Hard Winter.[40] One account from the history of Brown County, Nebraska, describes the conditions:

> The prairies were covered with snow so deep that the cattle could not graze on the buffalo grass on which the ranchers relied for their winter feed. The snow came early in the fall and laid on the ground all winter. It was so deep that the cattle could not travel, and at times a crust of ice covered the surface of it making travel impossible as the cattle sank into the snow and thousands of head starved to death, sometimes in sight of the hay which ranchers had put up to be fed when the cattle could not graze.[41]

Though vulnerable livestock did perish in October and throughout the winter, there are few documented human deaths from the elements. According to Doane Robinson, a "few human lives were lost in South Dakota" during the October blizzard, likely the blizzard that caught the most dwellers by surprise.[42] Other lives were lost along rail lines, largely related to attempts to clear the tracks, and at least a few tales of death due to exposure made the local newspapers.[43]

Though Wilder often wrote about events with uncanny accuracy, *The Long Winter* remains a work of fiction in which she took artistic license. For example, she typically portrayed blizzards as three-day events; a motif through the book is the cycle of three days of blizzards followed by one day of clearing, which provides rhythm and builds tension in the narrative. In reality, the snowstorms varied in duration, usually lasting for one or two days, followed by a day or two of blowing snow that limited visibility even when new snow

40. Cattle deaths for 1880 are noted in the *Forestberg/Miner County (Dakota Territory) Mercury*, 22 Oct. 1880, and the *Chatfield (MN) Democrat*, 22 Oct. 1880.

41. "Winter of 1880–81, Brown County, Nebraska," Nebraska Genealogy, accessed 27 Dec. 2023, https://nebraskagenealogy.com/brown/winter.htm.

42. Robinson, *History of South Dakota*, 1:303.

43. Wilson, *The Beautiful Snow*, pp. 117, 168, 250–51.

was no longer falling. Breaks between snows lasted anywhere from a few days to a week, with a couple notable periods of roughly two weeks in which snow fell more frequently.

Wilder frequently shifted, compressed, and extended timelines throughout her books to suit her purposes, so not every event in her life happened in the same sequence as she relayed it, nor can the duration between events be interpreted as literal. If a reader tried to backtrack from the Christmas blizzard to the school blizzard in December 1880, the dates are unlikely to match the records. They are, nonetheless, close enough to tell a compelling story that still captures the feel of the blizzards, their relentlessness, and their severity.

The Ingalls family, the town of De Smet, and families and communities like them survived the Hard Winter of 1880–1881. They survived even with minimal forecast information to prepare for the winter ahead or for any individual snowstorm or blizzard. They survived with depleted fuel and food, through cold and hunger. They survived despite being weakened, scared, hungry, and cold. They survived, but they were changed.

In *Little Town on the Prairie*, Pa takes particular care to prepare for another extreme winter in the fall of 1881. He applies the lessons he has learned about the family's vulnerability to disastrous winter weather, even though he cannot be sure that the coming winter will be as devastating as the last one. His range of experience has increased by just one data point, but an important one—the worst that winter can offer, as far as his experience goes. As it turned out, the winter of 1881–1882 was the fourteenth mildest on record as of 2023–2024, at least at nearby Huron, South Dakota.

We gather information about the possibilities of future hazards by looking at those hazards in the past, seeing the range of possibilities, and deciding how much risk we are willing to accept and how much effort, time, and money we are willing to spend to prepare for those extreme events. Science also allows us to look to the hazards of the future by understanding how the range of possibilities may change, how often those possibilities may repeat, and how precursor conditions set up extreme weather.

4

If Only We Understood

Perhaps the biggest hallmark of scientists is that it's hard to get them to agree. Scientists love to argue. They might call it debating, or publishing, or presenting. By nature, scientists are meticulous about details, even nitpicky. They ask each other pointed questions at conferences, continuing the conversations after hours over dinner or drinks, as they try to clarify each other's meanings, defend their results and positions, or dig into the questions raised by the science at hand. In the virtual world, they have long ongoing debates on social media, through email, and on conference calls.

Scientists often ask questions like toddlers trailing a parent around a room: "Why? How? What's this?" The answer to one question usually leads to a dozen new ones. Sometimes, scientists are not asking questions of each other, but of themselves. Asking myself just one scientific question about *The Long Winter* (Was it as bad as Wilder said it was?) started the cascade that resulted in this book:

How do scientists know how severe it was?
Why was it so severe?
Was there another winter as severe?
Could a "Hard Winter" happen again?

What would be the effects of a "Hard Winter" today?

Would scientists know in advance if it was coming? How?

What would be the advance signs of a long and difficult winter like the Hard Winter?

When scientists do agree, especially when many of them agree over a long stretch of time, the evidence must be so compelling that no other explanation satisfies. The alternatives have not held up to scrutiny by themselves and their peers; they are not scientifically sound, not viable. Agreement might start with just a few scientists, perhaps among collaborators. The scientific explanation gains the agreement of hundreds or even thousands of scientists as additional researchers follow the scientific method to reproduce and confirm its results.

The scientific method starts with a *hypothesis*, a possible explanation about the natural world that can and should be tested, worded in such a way that it can be proven incorrect. Among scientists, a hypothesis becomes a *theory* when it has been tested and corroborated repeatedly and is strongly substantiated. Non-scientists often use the word "theory" casually, such as having a "theory" about which animal is digging holes in the backyard. What they mean is that they have a *hypothesis*. To scientists, theories are what non-scientists might call "laws" or "facts"—things that are so well tested that scientists accept them as the best and perhaps only explanation. New information may test the theory, and if they aren't compatible, the theory is either adjusted or rejected. When a theory, such as gravity, evolution, climate change, relativity, or cosmic expansion, has held up for a long time to repeated testing, it becomes stronger and more widely accepted.

And that is how science earns trust: the scientific process filters out questionable findings and elevates the ones that last.

Define "Worst"

In the weather world, meteorologists often compare one weather event to another, or one season to another. They assign rankings to tornadoes with the enhanced Fujita scale, ranking them from EF0 (weakest) to EF5 (strongest) based on the damage they cause.[1] Hurricanes rank on the Saffir-Simpson scale from a Category 1 (weakest) to a Category 5 (strongest), based mostly on wind speed.[2] Even droughts are ranked by the U.S. Drought Monitor from D0 (abnormally dry) to D4 (extreme drought), based on input from a wide range of sources.[3]

The story of ranking the severity of an entire winter season starts here, with the Hard Winter of 1880–1881, and provides a context to understand not just the Hard Winter but all the winters with daily weather records. To determine whether the Hard Winter was the worst on record, first we must define "worst." My closest collaborator on this question has been Steve Hilberg, retired from the Midwestern Regional Climate Center. An all-in winter fan, Hilberg too had pondered how to classify winters as mild, severe, or something in between. As he helped me to understand, "worst" is too subjective a term in reference to our winter severity index. Some people love winter while others hate it. Some industries are shut down, but others thrive. Some animals and plants need occasional harsh winters for their health. We can't put a value judgment on weather and seasons, at least scientifically, because it doesn't apply universally.

Hilberg and I considered what the end result of calculating winter severity should look like: line graph charts, first with dips and rises like a roller coaster, and then an always-upward line with slower and faster rises like a growth chart. We talked about the character of winters, the factors that make them hard on people, towns, ecosystems, and industries.

Dubbing our index the Accumulated Winter Season Severity Index, or AWSSI (pronounced "Aussie"), our first major concern in the process of ranking a winter's severity was to determine which

1. Charles A. Doswell III, Harold E. Brooks, and Nikolai Dotzek, "On the Implementation of the Enhanced Fujita Scale in the USA," *Atmospheric Research* 93, nos. 1–3 (2009): 554–63.

2. Robert H. Simpson, "The Hurricane Disaster-Potential scale," *Weatherwise* 27, no. 4 (1974): 169–86.

3. Mark Svoboda, Doug LeComte, Mike Hayes, Richard Heim, Karin Gleason, Jim Angel, Brad Rippey et al., "The Drought Monitor," *Bulletin of the American Meteorological Society* 83, no. 8 (2002): 1181–90.

parameters to include. Certainly, snowfall was one, but we needed to be specific about what tied snowfall to severity. The amount of snow in one snowstorm matters, as does the frequency and the duration that the snow stays on the ground after it falls. Ice storms are dangerous and disruptive. Temperatures matter—cold days, colder nights, and how long the cold lasts affects everything from heating bills to the substances applied to roads to add traction. Wind creates winter havoc, mainly through wind chills and blowing and drifting snow.

We had to be able to calculate winter severity with widely available data, which narrowed the factors that we could consider. Documentation of daily high and low temperatures widely available at many sites for the length of weather records. Measurements of precipitation are widely available, too. But at many sites, snowfall and snow depth measurements began long after temperature and precipitation measurements. Wind is available from hourly records, but not from daily weather records, and wind chill is not readily available. There is no official source of information about ice storms, either; while several researchers have created their own records for their own studies, none are complete or continuous.[4] We opted to use daily high and low temperatures, snowfall, and snow depth in the AWSSI. We also chose to investigate whether we might be able to approximate winter severity from precipitation and temperatures without knowing the actual snowfall and snow depth measurement. Ice and wind could not make the cut, even if they have serious consequences.

From there, we had to make the leap to specifics, that is, specifically how to account for the temperatures, snowfall, and snow depth in a mathematical formula that would build the AWSSI.

Using ranges of temperatures, we set up a scoring chart. A high or low temperature between 25 and 32°F earns one point for the day, while temperatures of -20 or colder tally fifteen points, giving greater weight to those extreme values that are a danger to humans

4. Tamara G. Houston and Stanley A. Changnon, "Freezing Rain Events: A Major Weather Hazard in the Conterminous US," *Natural Hazards* 40 (2007): 485–94, https://doi.org/10.1007/s11069-006-9006-0.

and impede our lives. Using a range of values instead of specific values provides a little grace to records that may have been more imprecise at times due to poor siting, site moves, or differences in observation times.

Snowfall also earns a score based on thresholds. A daily total of 1.0 to 1.9 inches would score two points, while a more disruptive 6.0 to 6.9 inches runs up nine points, and an extreme of 30.0 inches or more would max out the daily score at forty-five points. As with temperatures, using ranges for points instead of precise values accounts for the difficulty of measuring snowfall precisely.

The final set of points are from the snow depth measurement—the snowpack on the ground, whether it is fresh or encrusted from a winter's worth of snow. While a snow depth of 3 inches gains three points, a snowpack of 12 to 14 inches adds seven points to each day.

Any winter day can add anywhere from zero points (if there is no snow and the temperature stays above freezing) to a spike of ten to twenty points or more. For example, if the high temperature was 15 degrees (three points), the low was -1 (seven points), a powdery 1.3 inches of snow fell (two points), and there was a snowpack of 3 inches on the ground (three points), the resulting total would be fifteen points. For a day with a high of 25 (one point), a low of 20 (two points), 5.5 inches of fresh snow (seven points), and a 6-inch snow depth (five points), the total would also be fifteen points. Though the days have a different character, the point totals are the same.

Next, we had to determine what defines the start and end to winter, so that the index had a defined beginning and end to the winter season. We polled our colleagues, friends, and family about what should define the start of winter, with a range of answers:

"The winter solstice (usually 21 December)"
"1 December"
"The first measurable snowfall"
"The first time I have to put on a coat"
"The first time I have to shovel snow"
"The first time temperatures fall below freezing"
"The first time temperatures stay below freezing"

Results for defining the end of winter were equally mixed:

"The spring equinox (usually 21 March)"
"1 March"
"1 April"
"The last freeze"
"The last snowfall"
"The last time I shovel snow"
"When snow piles are gone"
"When the daffodils (or other plants) emerge above ground"
"When I plant (a certain plant or flower) in my garden"

Hilberg and I chose a definition that accounted for both the calendar date and the weather conditions. While we decided that we couldn't scientifically account for such subjective factors as when one reaches for a coat or puts away a shovel, we also set parameters that reflected those opinions—that "putting away the shovel" means no more snow, for instance. Winter, for the AWSSI, begins when the first measurable snow falls or when the high temperature does not rise above freezing for the day, but no later than 1 December. Winter ends when the last measurable snow on the ground is gone, after the last measurable snow falls, or after the last day that stays below freezing, but no earlier than 1 March.

Each day's point total is added through the season for a running total, boosting the winter's total score with each day's winter conditions. The final AWSSI score puts a number on the winter, like the final score of a basketball game. The final score is marked by "runs" of faster point accumulation and "lulls" with little addition to the score—it is never a smooth upward climb to the final score.

At a given site, the scores over many years can be compared, their severity ranked by their point totals. Scores also can be compared between sites, both in absolute values (that is, one city has a more severe winter than another because its point total was higher) and in rankings or other relative values (that is, this was one of the ten most extreme winters in both cities, or among the top 20 percent of winters).

The final value can help put a number on the severity of the winter, but it does not necessarily define the character of that winter. A high score could come from a winter that was snowy, brutally cold, or both. Teasing apart the temperature and snowfall scores helps us better understand the character of the winter by understanding the contributions of each component to that total.

The final obstacle in the winter index problem was a substantial one: Snowfall records are relatively young compared to the length of temperature and precipitation records at most sites. No archived weather observing sites directly measured snowfall during the Hard Winter of 1880–1881, though several in the region kept temperature and precipitation records. We had to find a way to mathematically estimate the snowfall and snow depth based only on temperatures and precipitation. There is no precise and simple formula to convert temperature and precipitation amount to snowfall—the amount of snow depends on not just the temperature and humidity at ground level, but also thousands of feet up into the atmosphere, the duration of those temperatures and humidities, and even the wind. All of those factors can affect the type of snow crystals that fall, from icy needles to fluffy dendrites to shattered fragments, which in turn affects how the snow accumulates on the ground.

With no hope of gauging the temperatures at ground and atmospheric levels at every site for every day of its record, Hilberg and I did the best we could by leaning on the work of scientists before us, such as Charles Fisk, who devised a mathematical formula to do exactly what we wanted: estimate the snowfall based on only the daily temperature and precipitation readings, building his model from thousands of observations. His test case was, ironically, the Hard Winter of 1880–1881 in Minneapolis/St. Paul, Minnesota, and he had estimated that over 100 inches of snow fell that winter.[5] We started with his formula, refined it as we found some room for improvement, and incorporated it into a second version of the winter index.

In this second version, we had to account for snow depth. Here, we built our own mathematical formula. We started with the new snow that fell. We multiplied it by a known compaction factor, a

5. Charles J. Fisk, "A Multivariate Analysis of Summary-of-the-day Snowfall Statistics vs. Same-day Water Precipitation and Temperature Recordings," preprints, *American Meteorological Society 17th Conference on Applied Climatology*.

value that accounts for snow settling as it sits and evaporating, regardless of temperatures. Finally, when temperatures rise above freezing, we cut the snow totals by factors based on the temperature to account for melting.

The range of values that we chose to use for snow measurements served us well for the estimated-snowfall version of the AWSSI, because in addition to accounting for snow measurement challenges, it also gave us leeway for the differences between using measured snowfall and using a calculated estimate. We found that our calculation was within about 5 percent of the measured snowfall in most locations, which scientifically means that it's acceptable to use for relative comparisons of the index. In the interest of scientific honesty, we also discovered that the one place where the calculated version of the AWSSI did not behave so well was in the warmer and snowier eastern Great Lakes, in places like Buffalo, and Albany. It did, however, work very well in the Plains and Midwest, right where I wanted to apply it to study the Hard Winter of 1880–1881.

Hilberg and I, along with Martha Shulski and Kenneth Hubbard, who helped bring AWSSI to the finish line, published our work in the *Journal of Applied Meteorology and Climatology*, allowing our peers to review it for validity and to provide feedback before it enters into the body of scientific knowledge.[6] Since then, other researchers have validated the AWSSI and used it in their own research and applications as it gains acceptance in the scientific community. The Weather Channel later dubbed the AWSSI the "Winter Misery Index" as it gained some attention to put the hard winters of 2012–2013 and 2013–2014 into context.[7]

The AWSSI met our goal of providing a way to objectively compare the severity of winter seasons at a site or from site to site, based

6. Barbara E. Mayes Boustead, Steve D. Hilberg, Martha D. Shulski, and Kenneth G. Hubbard, "The Accumulated Winter Season Severity Index (AWSSI)," *Journal of Applied Meteorology and Climatology* 54, no. 8: 1693–1712, https://doi.org/10.1175/JAMC-D-14-0217.1.

7. Jonathan Erdman, "What 'Winter Misery'? It's the Record Mildest Start to the Season in More Than 20 U.S. Cities," The Weather Channel, 29 Dec. 2023, https://weather.com/science/weather-explainers/news/2023-12-29-winter-misery-index-mildest-start-december-2023.

on quantitative criteria. Even sites in milder climes, like Atlanta or Oklahoma City, can account for the relative severity of a winter compared to its own history, even if their tallies would scarcely register at sites like Minneapolis/St. Paul or Bismarck. Now housed on the Midwestern Regional Climate Center's website, the AWSSI is available each year, with the running tally available for over 100 sites across the country, updated each day.[8] The sites are color-coded by relative severity at their locations, from record mild to record extreme. The map and the snow totals that feed it allow for some friendly city-versus-city competitions, as well as a visual interpretation of the severity of the ongoing winter across the United States.

Ranking the Long Winter's Severity

By at least its perceived impacts, the Hard Winter is virtually unmatched. The ferocity of the Hard Winter—the unrelenting winds, biting cold, and overwhelming snowfalls—left deep scars imprinted in the history of the Dakotas, Nebraska, Iowa, and Minnesota. Having survived the winter became a badge of courage for the affected towns and their residents.

Perceptions and legends, however, can tend to grow out of proportion, leaving questions about the scientific backing behind the stories. The AWSSI, using the version that estimates snowfall and snow depth based on precipitation and temperature, allows us to rank the Hard Winter among all of those on record for a given site, providing data to quantify the stories and thus deepening our understanding of the experience.

By the calculations of AWSSI, the Hard Winter of 1880–1881 was indeed the most extreme winter on record for Omaha, Nebraska; Des Moines, Iowa; Helena and Havre, Montana; and Dodge City, Kansas. It was the second most extreme winter in Detroit, Michigan, and Cleveland, Ohio, and it ranks among the top five most extreme winters in many other places: Minneapolis/St. Paul, Minnesota; Madison and Milwaukee, Wisconsin, Denver, Colorado; New York City; Moline, Illinois, North Platte, Nebraska; Columbus,

8. AWSSI via the Midwestern Regional Climate Center, https://mrcc.purdue.edu/research/awssi.

Ohio; Lansing, Michigan; and St. Louis, Missouri, just to name a few.

Wilder's perceptions are supported by the data. The snowiness of the Hard Winter—snowfall and snow depth were both estimated using precipitation and temperature observations—is what made it so severe. The snow component of AWSSI ranks the Hard Winter as the most extreme at Minneapolis/St. Paul, Helena, Detroit, Madison, and Columbus. It ranks in the top five for the most extreme snowfall component of AWSSI at Des Moines, Moline, Havre, Milwaukee, and Dubuque, Iowa.[9]

Although snow was the more extreme contributor to the total AWSSI at many sites, temperatures also plunged far below normal. It was the most extreme winter on records for temperatures, by AWSSI calculations, for Helena, Des Moines, Omaha, and Dodge City (the same four sites that ranked as most extreme for the total AWSSI). It ranks in the top five for temperature at North Platte, Denver, Cleveland, New York City, Detroit, and St. Louis.

For almost all the continental United States, the Hard Winter ranks among the 20 percent most severe, with more moderate conditions extending up the East and West coasts. The winter was at its most extreme, compared to the history of all years in the data records, in a swath centered around Nebraska and adjacent areas of South Dakota, Iowa, and Minnesota (southwestern Kansas, around Dodge City, could also be included, but records for the Hard Winter are scant). Whether included in the swath or as a second epicenter, the winter also made its mark in Montana.

In other words, De Smet was in the heart of the Hard Winter, potentially in the worst of it.

Colder Climate

Underlying every other culprit in the severity of the Hard Winter of 1880–1881 was the base climatology in the 1880s, which was quite a bit colder than current conditions. Global, United States, and South Dakota temperatures have been rising notably from the late nine-

9. Barbara Mayes Boustead, Martha D. Shulski, and Steven D. Hilberg, "The Long Winter of 1880/81," *Bulletin of the American Meteorological Society* 101, no. 6 (2020): E797–E813, https://doi.org/10.1175/BAMS-D-19-0014.1.

teenth century through today, with no end in sight. Every fluctuation of weather and climate patterns in the 1880s moved around a colder background climate state.

Imagine a bathtub filled halfway with water. A child swings his arm back and forth over the surface of the water, making waves that are a couple of inches above the surface level of the water when it is flat. The child giggles and splashes, and the waves stay safely in the bathtub, contained by the sides of the tub. If the water is raised to just a couple inches from the top of the tub, the same movement from the child, producing the same waves, now causes them to splash over the edge. The child's force on the water did not change, only the water's starting level did. But because that starting level was different, the impact of the water was quite a bit different.

Similarly, a weather pattern that would cause the temperature to dip below normal will cause the temperature to drop to a colder reading if that "normal" is colder. Not every change in climate patterns is quite so direct; the way the patterns feed back on each other can amplify some changes and dampen others. But in general, a weather pattern that causes below-normal temperatures would have caused colder temperatures in the 1880s than it would in the 2020s.

The average winter temperature across South Dakota was about 3 to 5°F colder in the late 1800s than it is today.[10] This may not sound like much, but that shift in the average temperature alters the chance of extremes dramatically. When the average temperature is just 3° colder, the chance of temperatures below the threshold of the coldest 10 percent of the record leaps three- or fourfold. In other words, the atmosphere in 1880 was primed for colder extremes.

A colder background climate does not guarantee cold weather every year, and a warming climate doesn't mean every winter will be among the warmest on record; the winter of 1881–1882 ranks among the most mild on record (fourteenth warmest, to be specific) in Huron, South Dakota, while the winter of 2009–2010 is the

10. Rebecca Frankson, Kenneth E. Kunkel, Sarah M. Champion, David R. Easterling, Natalie A. Umphlett, and Crystal J. Stiles, "South Dakota State Climate Summary 2022," *NOAA Technical Report NESDIS 150-SD* (Silver Spring, MD: NOAA/NESDIS, 2022), https://statesummaries.ncics.org/chapter/sd.

twenty-seventh coldest on record. But a colder background climate makes those colder winters more likely to happen. Even excluding the Hard Winter, five of Huron's coldest fifteen winters are indeed from the 1880s.

Weak, Wavy Winds

The ocean temperatures and prevailing patterns halfway around the world can impact the weather of a landlocked location in the middle of the United States. Essentially, the atmosphere communicates with far-removed sources of energy. Energy from the ocean's heat is transferred through the atmosphere by the jet streams, the currents of faster winds that travel across the globe at an altitude of around 30,000 feet.[11] The polar jet stream channels its energy further north, while the subtropical jet stream ranges closer to the tropics. A fast, straight jet stream, sometimes called "zonal" flow, carries its weaker storm systems quickly from west to east. Weaker forcing, or buckling of the zonal jet streams, can lead to high-level winds that ripple and bend. These ripples translate into "troughs," or low-pressure areas, and "ridges," or high-pressure areas. The low-pressure areas are the ones that bring storminess, while high pressure areas tend to bring quiet weather. These rippling effects from faraway sources of forcing are dubbed "teleconnections" by meteorologists and climatologists (the prefix "tele-" means "at a distance" or "far off").[12]

One such teleconnection is the North Atlantic Oscillation.[13] There is a dominant area of high pressure near the Azores, off the coast of Spain, and a prevailing low-pressure area near Iceland. If the high is especially high, and the low is especially low, the height-

11. Toby N. Carlson, *Mid-Latitude Weather Systems* (American Meteorological Society, 1998), pp. 221–24; Howard B. Bluestein, *Synoptic-Dynamic Meteorology in Mid-Latitudes*, vol. II, *Observations and Theory of Weather Systems* (New York: Oxford University Press, 1993) pp. 130–31.

12. "Teleconnection," Glossary of Meteorology, American Meteorological Society, https://glossary.ametsoc.org/wiki/Teleconnection.

13. James W. Hurrell and Clara Deser, "North Atlantic Climate Variability: The Role of the North Atlantic Oscillation," *Journal of Marine Systems* 78, no. 1 (2009): 28–41, https://doi.org/10.1016j.jmarsys.2008.11.026.

ened difference in pressure supports a fast-moving and straighter jet stream, and it is dubbed a "positive" phase. When the high and low are not very strong, the weaker flow allows ripples to form in the jet stream. This "negative phase" often brings more low-pressure systems across North America in the winter, dumping colder air from Canada into the central and eastern United States. In a strongly negative phase, the persistent conditions have a powerful relationship with temperatures across most of the United States, bringing high confidence that temperatures will be colder than normal. Current science, however, does not yet allow forecasters to determine with much advance notice what phase of the North Atlantic Oscillation will dominate through the winter, so it is of limited value from a forecasting perspective. It is much more helpful in investigating winter seasons after they occur.

Much of this was not yet fully understood in 1880. Meteorologists in the late 1800s had only recently begun to use terms like "cyclone" for a storm system and "anticyclone" for a high-pressure system. The concepts of high and low pressure, the labeling of cold and warm fronts on maps, and their use in weather forecasting would not come about until introduced by Norwegian meteorologist Vilhelm Bjerknes in the 1910s to 1930s, long after the Hard Winter.[14] Weather maps drawn by meteorologists in the nascent U.S. Weather Bureau showed lines of equal pressure, but they did not label "L" for low pressure and "H" for high pressure, nor did they draw the blue-triangle cold fronts or red-semicircle warm fronts that are common on today's weather maps. Even the word "blizzard" was new in its use to describe a snowstorm, with mentions beginning in the 1870s and a sharp increase in use during and after the Hard Winter of 1880–1881.

The negative phase of the North Atlantic Oscillation in 1880–1881 was the fifth strongest on record from 1865 to 2024, and the January 1881 reading was the most negative January on record through 2024.[15] The observations support the hypothesis that the negative

14. Carlson, *Mid-Latitude Weather Systems*; Bluestein, *Synoptic-Dynamic Meteorology*.

15. North Atlantic Oscillation: https://climatedataguide.ucar.edu/climate-data/hurrell-north-atlantic-oscillation-nao-index-station-based.

phase of the North Atlantic Oscillation played a significant role in the severity of the Hard Winter.

The Boy Child

Halfway around the globe, in the vast Pacific Ocean near the equator, ocean temperatures rise above and fall below normal as the ocean sloshes east and west. The sloshing is driven by the trade winds that blow from east to west across the Pacific. Blow across the water in a bowl and you will see how your breath pushes it away from you. Stand beside the waters of a lake and you will feel that the water is warmer if the wind blows in to shore and colder if it blows out. Over the ocean, the wind has thousands of miles of fetch to push the waters around.

The blowing of the trade winds creates a pool of warmer waters in the western Pacific near the equator. When the trade winds are stronger than normal, they push even more of the warm surface water toward Indonesia, and cold water wells up from the deep off the west coast of South America. Tropical thunderstorms tend to follow the warmest water, driving the subtropical jet stream northward into a wavier pattern. This wavier jet stream pattern is better connected to cold Arctic air, dragging it out of Canada and into the northern United States. The southern Plains and Gulf Coast can be left high and dry as the jet stream pushes north, tipping the odds toward droughts and warmer temperatures. This colder phase is dubbed "La Niña."

The trade winds sometimes slacken. On rare occasions they can even reverse and blow from west to east; when this happens, the warm surface water sloshes back toward the east and pools off the coast of South America, or sometimes closer to Hawaii and the central Pacific Ocean. Thunderstorms continue to follow the warmer waters, but they drive the jet stream straighter, across Mexico, the Gulf of Mexico, and perhaps the Gulf Coast, and those areas may experience more storminess and cooler weather during the winter. But the subtropical jet stream remains disconnected from its polar cousin to the north, and often colder air struggles to reach the northern and central United States. Peruvian fishing communities, recognizing the weather pattern's devastating impact on their fish harvest, dubbed it "El Niño," or "boy child," for its tendency to develop around Christmas. La Niña earned the name as the opposite of El Niño during the 1900s. In between is a "neutral" pattern, when the ocean responds to average trade winds.[16]

The patterns create tendencies, not guarantees. Weaker El Niño and La Niña conditions tip the odds more weakly than their stronger counterparts. The signals of El Niño and La Niña can be masked by other patterns, or they can interact with other patterns in ways that change their impacts. Though they are not the strongest signals to affect the United States, they are the most predictable ones.[17] Forecasters can predict the presence of El Niño or La Niña with confidence months before it arrives, which is usually in the late fall or winter. Because forecasters know the temperature and precipitation tendencies of both patterns, they can use their prediction to tip the odds in the United States based on which pattern is emerging. Because other, less predictable patterns can interfere, like the North Atlantic Oscillation, for example, the forecasts are not perfect. But they are informed, based on what little information forecasters do have to tip the odds for seasons ahead.

Throughout the Hard Winter, weak El Niño conditions contributed to the weather patterns.[18] El Niño winters, taken by themselves, tend to bring a slight lean toward above-normal temperatures in the northern Plains, while La Niña winters are more typically associated with colder winter conditions. In fact, a winter season forecast in 1880 might have leaned toward either no tilt in the odds or a slight tilt toward above-normal temperatures, based on just the emerging weak El Niño. But when El Niño is coupled with a negative phase of the North Atlantic Oscillation, as it was in 1880–

16. There are a good number of scholarly papers on ENSO, but for a deeper read for interested non-technical readers, see Brian Fagan, *Floods, Famines, and Emperors: El Niño and the Fate of Civilizations* (New York: Basic Books, 1999).

17. Anthony G. Barnston and Robert E. Livezey, "Classification, Seasonality and Persistence of Low-Frequency Atmospheric Circulation Patterns," *Monthly Weather Review* 115, no. 6 (1987): 1083–1126, https://doi.org/10.1175/1520-0493(1987)115<1083:CSAPOL>2.0.CO;2.

18. Brian McNoldy, "Historic Oceanic Niño Index and Niño 3.4 Data," Rosenstiel School of Marine and Atmospheric Science, University of Miami, accessed 14 June 2019, http://bmcnoldy.rsmas.miami.edu/tropics/oni/.

1881, the combination leans toward being colder and wetter than normal.[19] Indeed, when this pattern repeated itself in 2009–2010, with a strongly negative North Atlantic Oscillation and a weak El Niño, cold and snowy conditions dominated the central and eastern United States from December through February, with three widespread blizzards in December. Snowfall in December 2009 broke records from Oklahoma to South Dakota to Washington, D.C., and Pennsylvania. On the background of a warmer climate, though, the winter of 2009–2010 did not last as long as the Hard Winter and did not match its overall severity, with a milder late winter and spring.

Other Long, Hard Winters

If the Hard Winter was not the most extreme winter at a location, then another one was. Another winter, also cold, also snowy, also long. Surely, a winter so extreme would have notable impacts of its own, stories that reach beyond the communities and the people who were directly affected.

Hard winters still occur in the United States, though not as frequently in any one location as they did a century ago. Even in a warming climate, it is possible to have record or near-record extreme winters, as weather patterns can still converge to produce extreme cold and snow. Depending on where a person lives in the United States, an extreme winter may have occurred as recently as within the last decade, or as long ago as the 1970s or 1950s.

From southern Michigan to the Ohio Valley and into Missouri, the winters of the late 1970s dealt brutal blows of winter conditions, sometimes in consecutive years. Both the winters of 1977–1978 and 1978–1979 stand as the most extreme on record in locations from Cincinnati, Ohio, to St. Louis, Missouri, to North Platte, Nebraska. The winters were particularly snowy in the region, with several deep snowfalls during the season and some of the strongest wintertime storm systems known to modern weather records. The "Cleveland Superbomb" storm of 27–29 January 1978, one of many powerful storms that winter, brought record low air pressure and a

19. Barbara E. Boustead, "The Hard Winter of 1880–1881: Climatological Context and Communication via a Laura Ingalls Wilder Narrative" (Ph.D. diss., School of Natural Resources, University of Nebraska–Lincoln, 2014), pp. 92–95.

true blizzard that pushed the one to four feet or more of fresh snow into drifts that reached the tops of houses.[20] The worst of the blizzard struck Michigan and adjoining areas across the Great Lakes and Ontario, Canada, causing dozens of deaths and clogging travel for days.

I was just a baby when the blizzard of 1978 struck, but I have vivid memories of the pictures in our old family albums, now lost to the ages. Drifts of snow reached the top of our mobile home, the front window peeking through like an open eye peering at the whipped white frosting, framed by gold siding. Snow drifted over the hood and roof of our family's early-1970s bulky sedan, with just the rear tailgate left exposed in shades of blue.

A modern hard winter brought near-record to record extreme AWSSI rankings to parts of the Great Lakes and upper Midwest in 2013–2014. Many locations in Michigan's lower peninsula, as well as a handful in Wisconsin and Minnesota, experienced a record extreme of winter severity, exceeding the previous record winters of the late 1970s. The severity of a winter like 2013–2014 and those late 1970s winters now occurs around once per generation (every 30–40 years or so), when it used to occur more like once every ten years or so a hundred years before. In 2013–2014, both cold and snowfall drove the winter's severity. Researchers concluded that the winter of 2013–2014 is now a once-in-a-century event, while it would have been a once-in-a-decade event at the end of the 1800s.[21] The winter conditions impeded transportation and cost billions to the economy of the affected regions.

Across part of the upper Midwest, from southern Michigan

20. Gregory J. Hakim, Lance F. Bosart, and Daniel Keyser, "The Ohio Valley Wave-Merger Cyclogenesis Event of 25–26 January 1978. Part I: Multiscale Case Study," *Monthly Weather Review* 123, no. 9 (1995): 2663–92, https://doi.org/10.1175/1520-0493(1995)123<2663: TOVWMC>2.0.CO;2.

21. Klaus Wolter, Martin Hoerling, Jon K. Eischeid, Geert Jan Van Oldenborgh, Xiao-Wei Quan, John E. Walsh, Thomas N. Chase, and Randall M. Dole, "How Unusual was the Cold Winter of 2013–14 in the Upper Midwest?," *Bulletin of the American Meteorological Society* 96, no. 12 (2015), S10–S14, https://doi.org/10.1175/BAMS-D-15-00126.1; also cited in "How Unusual was the Long, Cold Winter of 2013–14?," NOAA, 5 Nov. 2015, https://psl.noaa.gov/news/2015/110515.html.

through Wisconsin and Minnesota and into North Dakota, the winter of 1874–1875 ranks as the most severe on record. The swath includes Detroit, Milwaukee, Madison, La Crosse, Minneapolis/St. Paul, Duluth, and Bismarck. It is notable that the Ingalls family lived right in the middle of that swath in Walnut Grove, Minnesota, during that winter and experienced those severe conditions. Yet the winter is not marked as exceptionally long and hard in Wilder's stories. She pulls a few stories of blizzards, including a story of Pa caught out in the elements through an entire blizzard before Christmas, and she mentions cold temperatures in *On the Banks of Plum Creek*. A closer look at this "other hard winter" lies ahead in this book. But Wilder did not dedicate an entire book to it, nor did she label it with gripping names.

The extreme winter may have received diminished attention in Wilder's narrative for a number of reasons. In *On the Banks of Plum Creek*, the locust ("grasshopper") infestation drives the narrative. With emphasis on mild weather contributing to locust spread, perhaps the story of the hard winter in the middle of the locust infestation simply did not fit the narrative. Also, the town of Walnut Grove was established by several decades by the time the winter of 1874–1875 struck, as were many of the settlers in the area. The trains provided steady and reliable service from Minneapolis/St. Paul, and many provisions came from local sources. Trees provided wood for fuel, and more established farms ensured a somewhat steadier food supply. Perhaps the community was better able to absorb strained fuel and food sources, making families like the Ingalls with stretched resources less vulnerable to the strain. Finally, it was temperatures that drove the severity of the winter in 1874–1875, and perhaps being extra cold did not have the same brutality as being extra snowy. Cold does not stop trains like snow does. Cold does not make travel impossible like snow does. An exceptionally cold winter, by Minnesota standards, is a colder version of something that is already very cold. Maybe, for the Ingalls family and the Walnut Grove community, snow was a greater threat than cold.

When scientists look for the cause of an event—a disaster, a tragedy, a one-off wreck or a systematic failure—their process includes a root cause analysis. Scientists start at the outcome and work their way backward, asking "why" as many times as they can until they find the origins of issues that contributed to the tragedy.[22] There is never just one cause, even in disasters that seem straightforward. Many factors converge to create a disaster.

For example, consider a building that collapses in a tornado, killing and injuring people inside. What caused those casualties? It was not just the tornado. It also was not only the construction of the building. Both of those are contributing factors, but scientists dig deeper. Was there a tornado warning? If so, how far in advance of the tornado was it issued? Did the people inside know that there was a tornado warning, and if so, when? Did they immediately seek the safest possible shelter, or were they delayed by not knowing where that was, confusion, mixed messages, or other indecision? Did they have enough advance notice to reach that shelter? Where were the failure points in the building's construction? Was the building construction up to code, and was the building well maintained?

If science is defined by any two words, those words would be "asking questions." Questions often lead to more questions, rather than to clear yes-or-no answers. Scientists form hypotheses to test answers to their questions, looping closer and closer to finding the most reasonable answers. They adjust when new information arrives, test it, incorporate it, and improve their understanding.

Science is not black and white. It is shades of gray and rainbows of colors, sometimes bright and sometimes as dim as a candle in a blizzard. It is probabilities and chances, rather than absolutes. It builds on decades and sometimes centuries of past work. It continually reassesses what we know. It leads us closer and closer to having the best explanation for our world—the most supported by observations and research, the closest to the measurable and observable data supporting it.

Science seeks a preponderance of evidence, and only rarely is it beyond the shadow of a doubt.

The questions about what may have caused the severity of the

22. Paul F. Wilson, Larry D. Dell, and Gaylord F. Anderson, *Root Cause Analysis: A Tool for Total Quality Management* (Milwaukee: ASQC Quality Press, 1993).

Hard Winter of 1880–1881 lead to layers of potential contributors to the hardships experienced by the Ingalls family and the town of De Smet, along with many families and communities like them. The baseline of a colder climate, the wave patterns in the atmosphere set up by the strongly negative North Atlantic Oscillation, the nudge from El Niño, and the randomness of weather itself all are among the factors and likely are not the only factors that produced deep snows and prolonged cold. The newness and poverty of many homesteaders exposed them to the elements, with already tenuous fuel and food supplies cut off for months. The human and natural elements converged into the worst of combinations.

None of the factors is the "smoking gun," beyond-the-shadow-of-a-doubt culprit in the Hard Winter story. Each of them has a clear and measurable role, though.

From the records and stories of all of the individual storms during the Hard Winter of 1880–1881, to understanding the weather patterns that contributed to the extreme winter severity, to comprehending just how extreme that winter was when compared to others, we have a clearer picture of just how the Hard Winter caused such hardship for the Ingalls family and the community of De Smet and others like it. The puzzle pieces include archived weather observations, Indigenous accounts, computer models, and an index that ranks the winters by severity. It also includes Wilder's own words, which provide insight into not just the numbers of the winter, but more importantly, the feeling of it and the effects of it.

Without Wilder's account and the words of other storytellers, the numbers lose their meaning and become merely another skeleton of past extreme weather. The Hard Winter of 1880–1881 would fade into the background with other winters that rank as extreme by the numbers. Her words let her readers, along with future scientists and researchers, distinguish the winters by character. She adds personality, color, character, and sound to the bones of the Hard Winter.

Winters like the doozies of the 1870s and 1880s used to happen once or twice per decade. By the mid-1900s, they occurred once or twice per generation. By the twenty-first century, the most extreme winters are once in a generation to once in a lifetime events.

Extreme cold and snow still can and will occur, but their grip is shorter and the space between them continues to grow. Something that was part of the lived experience of our communities and its people in the generation of our grandparents and great-grandparents is relegated to stories for our children, something they may get one or two chances to experience, and maybe something they can only relay in stories to their grandchildren as the events start to skip their lifetimes.

II

Pay Attention to Weather All Around Us

5

The Severe Cold of the North

In the northern climates where the Ingalls family spent most of Wilder's childhood, the cold was inescapable, inevitable, and nearly intolerable. Even in the warmer months, chill crept into their home, which lacked the modern comforts we take for granted, such as good insulation or being able to set the thermostat to a cozy 70 degrees. In the depths of winter, the kitchen stove might as well have been a candle compared to the frigid cold encroaching on the house. At times, they could see their breath indoors, indicating that the indoors temperature was around 45 degrees or colder. Water inside the house froze at night and had to be thawed in the morning.

For months at a time during the longer winters, the house felt like little more than an ice palace that blocked the wind from entering and the snow from falling on their heads. As we have seen, their claim shanty didn't even accomplish that during the October blizzard. Balanced on the edge of freezing throughout the Hard Winter, they survived the cold in other winters, too, enduring the unending knots in their stomachs that accompanied a prolonged fight against bitter temperatures. Sometimes the cold surprised them, but even when they had anticipated it and prepared, the cold could take their breath or their crops away.

"Sugar Snow" and Maple Syrup

While many of the trees in the Big Woods of Wisconsin had been cleared for farmland by the time Wilder was born, patches of mixed forest, with deciduous trees like sugar maple, elm, and oak, remained (and still remain) along the Mississippi River Valley.[1] During Wilder's years in Wisconsin (1867–1869 and 1871–1874), the sugar maples provided an abundant, if labor-intensive, source of sugar for local farmers.

In *Little House in the Big Woods*, Wilder describes a rather typical-sounding winter thaw, with the snow cover receding, the ice melting, and the bare ground expanding. Just as five-year-old Laura is ready to play outside, a nighttime snow blankets the ground once again. Pa, calling it a "sugar snow," rushes to help his parents, Grandma and Grandpa Ingalls, tap and process maple sap into sugar. Laura tastes the snow itself; determining it is not at all sweet, she wonders why it is a sugar snow. Not until he returns from his work that evening does Pa have time to explain the maple sugaring process to Laura and Mary: "It's called a sugar snow, because a snow this time of year means that men can make more sugar. You see, this little cold spell and the snow will hold back the leafing of the trees, and that makes a longer run of sap."[2] The labor-intensive process included racing around the tapped trees to empty buckets into a larger collection barrel, then taking that barrel to an even larger kettle over a fire to boil down the sap and concentrate its sugars. Depending on the temperature to which it was cooked, the sap would cool into syrup or granulated sugar. A week of such labor culminates in a vibrant sugaring-off party at the grandparents' home, with dancing, fiddling, and plenty of maple-sweetened treats. Young readers learn that pouring warmed maple syrup into a pan of snow results in candy, inspiring countless at-home maple-candy-making attempts since the book's publication.

1. G. Cottam and O. L. Loucks, *Early Vegetation of Wisconsin*, Wisconsin Geological and Natural History Survey, 1965, 8.5 x 11 in., http://wgnhs.uwex.edu/pubs/m035/.

2. Laura Ingalls Wilder, *Little House in the Big Woods*, (1932; New York: Harper Trophy, 1971), p. 119.

After maple trees endure a frozen winter, their sap begins to run as temperatures during the day rise above freezing, while still falling below freezing at night. The repeated freezes and thaws create daily cycles of positive and negative pressure inside the stems of the trees.[3] Daytime warming causes gases in the tree to expand in the stem, drawing sap upward through the trunk and branches to fill the space. Nighttime cooling causes the gases in the stem to compress as the water in the stem freezes, and the sap retreats toward the roots. Maple trees will produce sap for longer if the period of above-freezing daytime temperatures and below-freezing nighttime temperatures is prolonged. Tapping the tree allows the sap to flow out and be collected. The pressure difference is not unlike springing a mostly empty bottle of shampoo to its original shape, allowing the internal volume to return to normal and transforming the positive pressure of the collapsed sides into negative pressure inside that sends the contents toward the opening.

Sap flow is maximized during the time of year when high temperatures are mostly above freezing and low temperatures are mostly below freezing, namely, the transition from winter to spring. A sugar snow, while not itself causing the sap to run, signals that the cooler nighttime temperatures are not yet done for the season and that sap will continue to run.

Between 1872 and 1900, at La Crosse, Wisconsin (the nearest weather observing site to Pepin), high temperatures, on average, started to cross the freezing mark around 27–29 February. Low temperatures, on average, started to exceed freezing around 5–6 April. The length of actual overlap of these ideal conditions in any given year can fluctuate wildly, and there is not a smooth beginning or end point to the season during which high temperatures are above freezing and low temperatures are below freezing. As spring overtakes winter, gaps of several days to a week or more in between freezing nights can occur. Sap run seasons are optimal when below-

3. For more explanation of the positive/negative pressure effect, *see* Kristen Patino, "Tapping for Sugar Maple Sap on Campus," Washington University in St. Louis, 28 Feb. 2018, https://sustainability.wustl.edu/tapping-for-sugar-maple-sap-on-campus/.

freezing lows are more or less consecutive, while high temperatures remain above freezing.[4]

One growing degree day, with a base of 50°F, is accumulated when the average daily temperature (maximum + minimum / 2) is one degree above 50°F. Sap collection typically ends when trees reach bud break stage, which occurs around when the location has hit fifty-four growing degree days. In the late 1800s, the average date of bud break, or reaching fifty-four growing degree days, was 23 April; by the early twenty-first century, it had slid all the way up to 16 April.

Wilder did not make note of when the party at Grandma and Grandpa's house occurred. By the sequence of her *Pioneer Girl* manuscript, which itself contained anachronisms, Laura would have been about four years old, placing it in 1871. The Ingalls family, however, was just returning from Kansas in 1871 and likely was not back in time for the spring thaw. It is also unlikely that the dance occurred in 1874, when the Ingalls family departed Pepin. This leaves the springs of 1872 and 1873 as candidates for the sugar snow episode. Records indicate potential snows during the spring of both years.

If the dance occurred in 1872, it likely followed a return to cold temperatures and snowfall around 31 March. On that day, with a high temperature of just 35°F, a low of 28°F, and 0.50 inches of liquid-equivalent precipitation in Madison, Wisconsin, it is possible that several inches of snow fell. Through the preceding week, temperatures had risen into the mid 30s to around 40 degrees by day, with lows in the 20s to 32°F. Those temperatures would support sap flow, and a snowfall could have contributed to cooler temperatures until the snow cover melted. The sugaring weather would not last much longer, as temperatures were above freezing for several mornings in a row beginning on 6 April.

4. John Newell, "Revisiting the End of Maple Sugaring Season—Based on Growing Degree Days," Primal Woods Sugarers, accessed 11 Mar. 2024.; Kelly Cammerota and Aaron Lefland, "Maple Sugaring: The Sap Run," Land's Sake Farm, last modified 11 May 2017, https://landssake.org/maple-sugaring-the-sap-run/; and Melvin Tyree, "Maple Sap Exudation: How It Happens," *Maple Syrup Journal* 4, no. 1 (January 1984): 10–11.

If the sugar snow occurred in 1873, the date is harder to discern. On several days in late March, cold temperatures and precipitation were observed at one of the three nearby observing sites, but on no days did all three sites experience cold temperatures and precipitation. At La Crosse, temperatures on 24–25 March fell from 32 to 13°F, with a total of 0.07 inches of precipitation (likely snow) through the two days, translating to probably less than two inches of snow and more likely closer to an inch. Cold temperatures and very light precipitation also occurred at Minneapolis/St. Paul on 24–25 March, with 0.07 inches of precipitation (likely snow) and temperatures well below freezing. Additionally, on 28 March, with a high temperature of 36°F and a low of 21, 0.60 inches of precipitation fell, which could have fallen as rain, snow, or a rain-snow mix. Of the three sites, Minneapolis/St. Paul was the only one that recorded precipitation on 26–29 March. A little farther east, in Madison, temperatures on 26 March only rose to 23°F, with a low of 5; the 0.30 inches of precipitation would have fallen as snow, likely around 3 to 5 inches. Looking at 8–9 April, with high temperatures of 34–39°F and low temperatures of 25–30°F at the three nearby sites, it is difficult to determine if the 0.16–0.30 inches of precipitation fell as rain, a rain-snow mix, or snow.

Given the evidence for both 1872 and 1873, we can deduce that the "sugar snow" and run of sap most likely occurred in 1872. In addition to the potential snow around 31 March, temperatures above freezing by day and below freezing by night lingered for a week into April, which was not matched in 1873. The longer duration of the optimum temperatures would have supported a prolonged sap run, filling Grandpa and Grandma's buckets for a week into April. The sugaring-off dance may have occurred on Friday, 5 April, or Saturday, 6 April, near the tail end of the sap run.

July Freeze

Farmers and growers fear the sting of a late freeze and the damage it can do to crops, gardens, and fruit trees. The calendar date of a freeze is not nearly as important as the extent to which the crops have grown, the fruit tree buds have emerged, or the planting of delicate seeds has progressed. As winter turns to spring, the warming temperatures accumulate as growing degree days, during which plants come to life, stretch their tender new leaves and shoots, and tentatively send roots into the ground. Even an on-time freeze can cause damage if enough growing degree days have accumulated before it.

In *Farmer Boy*, the one book among the Little House series focused on Almanzo Wilder and his family, weather threatens the Wilder family's crops on their farm in Malone, New York. Corn growth lags behind Father's expectations into early summer, and his concern for the crop mounts. On the morning of 4 July, the Wilder parents rouse young Almanzo and his siblings to pour buckets of water over the corn, hoping to save it from a freeze in the pre-dawn hours. They labor for hours, watering about half the field before dawn, at Father's insistence that any plant not watered during the freezing temperatures would die when the sun touched it.

Even in the cool and snowy climate of upstate New York, even in the colder climate of the mid to late 1800s, true freeze conditions in July are exceedingly rare. Among nearby long-term weather stations in Burlington, Vermont, and Rochester and Buffalo, New York, the records extend as far back as 1872, with no freezes noted in the month of July. But temperatures in upstate New York and Vermont do rarely fall into the upper 30s in July—less frequently in the modern climate than in decades past, but still in the range of possibility. The temperature fell to 39°F in Burlington on 6 July 1962, and it has fallen into the 30s in Jamestown, New York, eight times since the 1920s. Those temperatures in the upper 30s can come with frost in low-lying areas on calm, clear nights. In the higher terrain of the Adirondack Mountains in eastern New York, temperatures at places like Saranac Lake and Newcomb fall into the 30s more frequently, and there are a few subfreezing July readings on record. These locations are not representative of Malone, though, because they are so much higher in elevation.

The scant observations available show that July of every year between 1866 and 1869—Almanzo Wilder was born in either 1857 or 1859—was mild, with temperatures as low as the 50s to 60s.[5] Going

5. Almanzo Wilder's birth date is a matter of some controversy among Wilder scholars. While 1857 is more widely cited, evidence suggests that he may have

back a little further in the record, one finds that on the morning of 4 July 1859, temperatures at 7:00 a.m. were as cold as 48°F in Burlington, 52°F in Rochester, and 39°F at Mount Morris (south of Rochester and east of Buffalo). Though these temperatures also seem far from freezing, the weather observation at Mount Morris included the note, "a frost in the lowlands!!"[6] Nathaniel Arbuckle, in rural Delaware County in southeast New York, also noted the frost in his diary: "This is the Glorious old forth [*sic*] the Cannon is moaning all around us we had quite a smart frost last night Something I niver saw before on the forth."[7] Both Delaware County and Mount Morris are located well south of Malone; Delaware County is nestled in the Adirondacks, while Mount Morris is in a river valley.

The chill of the morning of 4 July 1859 found its way into newspapers and journals from New York to Kentucky. From Rockland County, New York: "A clear and bright sky graces the dawn. Cool weather. People wear overcoats and farmers hoe corn while wearing heavy clothing."[8] From Athol, Massachusetts: "[A monument] dedicated on a bitter cold July 4, 1859."[9] From Bradford County, Pennsylvania: "killing frost every month in the year, in some places there was a flurry of snow on the 4th of July and so cold that persons wore overcoats at the celebrations."[10] From Pleasant Grove,

---

been born in 1859 and lied about his age when he filed his homestead claim in 1879 so that he would be eligible. *See* Fraser, *Prairie Fires*, p. 113.

6. Brian Belanger, "Historical Weather Events in Rockland County, New York from 1709 to 1909," Rockland Genealogy, last revised 12 Aug. 2012, rocklandgenealogy.org/Weather-RocklandCo-1709-1909.pdf.

7. Oneonta Arbuckle Index: http://www.oneonta.edu/library/dailylife/business/arbuckleindex.html; http://www.dcha-ny.org/researchingatdcha201810.pdf.

8. Belanger, "Historical Weather Events in Rockland County, New York."

9. George Barnes, *Telegram & Gazette* staff, "150-year-old mystery in Athol may be revealed: Town's Own 'National Treasure'," 16 Nov. 2009, http://atholhistoricalsociety.weebly.com/hapgood.html.

10. Clement F. Heverley, *History and Geography of Bradford County, Pennsylvania, 1615–1924* (Towanda, Pa.: Bradford County Historical Society, 1926), p. 573, https://digital.libraries.psu.edu/digital/collection/digitalbks2/id/9405.

Indiana: "July 4, 1859, frost."[11] From Lewis Collins and Richard H. Collins's *History of Kentucky*: "July 4, 1859, was so cold and inclement that picnic parties were compelled to return home for heavy wrappings."[12]

Given the records and stories from the region surrounding Malone, it is not a stretch to conclude that a frost or even a freeze occurred in Malone in the early morning of 4 July 1859.[13] Almanzo Wilder, either an infant or a toddler at the time, probably had little or no recollection of the event himself, but his parents and three older siblings would have remembered it and told him the story as part of their family history.

Without the option to move plants to a warm place to protect them, or to put a large cover over a field of crops, modern farmers must bring some scientifically backed measures of protection to their fields. Water remains one of the methods of choice, but the key is to apply it continuously, not just once.[14] As water freezes, it releases a small amount of heat energy, known as latent heat. The tiny amount of heat can be enough to keep the temperature at 32°F very close to the plant, but it only works as long as the temperatures are around 23–24°F or greater and only if the winds are relatively light, allowing the latent heat to remain close to the plant. Additionally, the application of water must be constant so that it can continue to freeze and release heat until temperatures rise above freezing. Many famous pictures of freezes in citrus-growing regions show trees laden with icicles; the icing is a result of spraying water to offset the freeze. As threatening as the icicles look, they are a visual reminder of the methods that might have kept the fruit safe.

If a freeze occurs on a calm, still morning, the subfreezing temperatures may be confined very close to the ground in a process

11. Richard C. Schmal, "Pioneer History: The Bryant Family, from the April 30, 1980, *Lowell Tribune*," 18 Sept. 2006, http://www.lowellpl.lib.in.us/s1980apr.htm.

12. Lewis Collins and Richard H. Collins, *History of Kentucky*, vol. 1 (Covington, Ky: Collins & Company, 1874), p. 245a.

13. 20th C v3 reanalysis of temps 4 July 1859.

14. Mark Longstrom, "Using Sprinklers to Protect Plants from Spring Freezes," Michigan State University Extension, 19 Mar. 2012, https://www.canr.msu.edu/news/using_sprinklers_to_protect_plants_from_spring_freezes.

known as radiational cooling. Using large fans to move the air allows it to mix with warmer air and keep temperatures above freezing at plant level. Likewise, during calm-wind freezes, growers can light fires in pits distributed through their orchards and fields. The fires both warm the air and create air movement, which is useful where setting up and powering a fan is impractical.

Regrettably, few of the methods above were known or available to the Wilder family in the 1850s. While the race to douse the crops with water builds narrative suspense in *Farmer Boy*, the practice the Wilders applied might have done more harm than good. The tiny amount of latent heat released on the initial freezing of a single bucket of water would not last until the day's warming brought temperatures above freezing. Also, as water evaporates off the ice (because it isn't arriving in a continuous sprinkle), the temperature near the ice falls colder than the air temperature. Rather than stopping at dawn, the Wilder family would have needed to keep pouring buckets over and over, right through dawn and until the temperatures rose above freezing and the ice began melting on its own. They were fortunate that temperatures were only a little below freezing.

The Other Hard Winter

The Hard Winter was not the only extreme winter of Laura Ingalls Wilder's childhood. During the winter of 1874–1875, when the Ingalls family was nestled along the banks of Plum Creek, record cold temperatures extended in a swath across Minnesota and Wisconsin. Although there was much less snowfall than during the Hard Winter, the winter of 1874–1875 ranks as the most extreme winter on record due to temperature. While Wilder mentions a couple of snows and blizzards in *On the Banks of Plum Creek*, she gives the winter only a passing mention as a welcome break that would loosen the grip of grasshoppers.

Indeed, that winter fell in the midst of the Rocky Mountain locust infestation. The narrative arc Wilder uses in *On the Banks of Plum Creek* to describe "grasshopper weather" includes mild winters between the hot and dry summers, even though weather records decisively indicate that this was not the case in Minnesota and surrounding parts of the upper Midwest and northern Plains. Wilder does build to a more severe winter near the end of the novel. It is plausible that the experiences of the Ingalls family during the "other" hard winter are relayed out of chronological order, like pictures arranged in a scrapbook to be more pleasing to the eye.

In one such story, Pa and Ma are away in town when a blizzard strikes. Laura, home alone with Mary and Carrie, is frightened by stories of children at home not having enough fuel to outlast a blizzard and freezing to death. At Laura's urging, she and Mary bring in an entire pile of cut wood, finishing just in time for Pa and Ma to return. Pa builds their wood-frame home in the spring of 1875, which is the backdrop for the woodpile story.[15]

While a heavy snowstorm or blizzard can strike at any time of year, it is more likely to have caught the Ingalls family by surprise if it was not in the heart of winter—that is, in either the fall or the spring. For example, on 25–26 October 1875, temperatures at Minneapolis/St. Paul fell from the 40s to around 30 degrees. During that time, 1.04 inches of liquid equivalent (rain and snow melted down together) of precipitation fell. No record remains of whether it fell as rain or snow—or how much—but the record from Fort Snelling does indicate mixed precipitation on 25 October and snow on 26 October, along with winds switching from southerly to northwesterly and increasing in speed as a cold front passed through the area. Such an event, with temperatures falling, precipitation changing to snow, and winds becoming northwesterly and increasing, could be the kind that startles a family, even if it is not an instant blizzard when it starts.

Though the Ingalls home was only a mile or two from town, being caught in the weather during a trip to town becomes something of a motif, as Pa is caught again returning home from town just before Christmas. This time, he does not make it home, riding out the storm in a sheltered snowbank after losing his way. Earlier drafts relayed the story of an unnamed man caught outside in a blizzard who perishes of exposure, then assigned the story to a neighbor character, and only in later drafts attributed the event to Pa, who

15. Wilder, *Pioneer Girl: The Annotated Autobiography*, p. 73.

survives.[16] Thus, it is plausible that the story did not occur, or at least not to Pa, and was added to demonstrate what *could* happen in the extreme winter conditions rather than as an event that did happen. If Pa truly had been caught out in a storm that lasted three to four days, it would have been fortunate (or miraculous) that he survived at all. Even if he was sheltered from the wind, he would still be exposed to the cold and relatively stationary, unable to move around and stay warm. Despite his new buffalo coat, he likely would have experienced hypothermia and could have died after several hours. His chances of surviving several days without warmth or movement would have been exceedingly low.[17]

Weather observations from Minneapolis/St. Paul do support the potential for a significant snowstorm on 16 December 1874, with plunging temperatures from 15–17 December, along with a half inch of liquid equivalent of precipitation that likely translated to six to eight inches of fresh snow. Because the temperatures reached as high as 48 degrees on 15 December, it is possible that residents like Pa thought it would be a nicer day before the cold temperatures and snow arrived. Fort Snelling's wind reports (direction and by force according to the Beaufort force equivalent) indicate that the winds out of the south increased from force 4 to force 6 on 15 December. Force 4 winds are equivalent to around 13–18 mph of sustained wind speed, likely with higher gusts, while force 6 winds are equivalent to roughly 26–31 mph sustained. Winds came around to northwesterly on 16 December as temperatures fell, with speeds at force 2 to force 4 through the day. Though a little less likely, another light snow did fall from 22–24 December 1874, with a total of just 0.08 inches of liquid equivalent of precipitation that may have translated to one to two inches of snow. Fort Snelling data indicate winds on 23 December of force 4 to force 6 out of the west to southwest, relatively strong compared to surrounding days.

Not to be outdone, snow also fell on Christmas Eve 1875. Temperatures dropped through the day and into Christmas Day, and

16. Wilder, *Pioneer Girl: The Revised Texts*, pp. 92–93.

17. Thanks to several colleagues of author Wendy Corsi Staub, who relayed their research related to writing crime thrillers.

the 0.57 inches of liquid equivalent of precipitation likely fell as another six to nine inches of snowfall. Winds at Fort Snelling rose from force 4 to force 6 out of the west on 24 December. Feasibly, either of the snow events in mid-December 1874 or Christmas Eve 1875 could have been the one that caught Pa or settlers like him in the open. These one-day storms would have been survivable, and it is far more likely that a person would survive a span of hours in a snowbank than three or four days, as Wilder's story would relate.

Nonetheless, the winter of 1874–1875, cold as it was, simply did not have the same impact on the Ingalls family as did the Hard Winter of 1880–1881. We can only guess why it was not a stronger piece of the family narrative. Perhaps the impact of the locusts in the summers of 1874 and 1875 simply overshadowed that of the winter between them, delivering a greater blow to the family and its stability. Including a hard winter might have detracted from the narrative focus on the grasshoppers and their effects. The severe and unusual snowfall of the Hard Winter may have been more memorable than a winter more characterized by severe and unusual cold temperatures. Another explanation is that Wilder, because she was younger in 1874–1875, did not pick up as much of the nuance of adult concerns about food and fuel. Or, as hard as it is to imagine after the losses of the first locust infestation, perhaps the family was not as vulnerable to a hard winter while they lived near Walnut Grove. They may have had enough fuel, with a more abundant supply of wood from trees near Plum Creek. The town itself was longer established, and therefore neighbors had more resources, unlike the first-year settlers that populated De Smet during its Hard Winter. Whatever the reasons, the family was not impacted enough to warrant more than a few chapters of stories about enduring winter conditions.

"Cold Miserable" Journeys

Sometimes, the cold is more of a mood than a measurement—more about the storytelling than the records.

"It was a cold miserable little journey and we were glad when we drove into Burr Oak," Wilder wrote in her *Pioneer Girl* manuscript.[18]

18. Wilder, *Pioneer Girl: The Annotated Autobiography*, p. 99.

She left readers with little information about her family's time there, relegating it to a short chapter of *Pioneer Girl.* It is entirely absent from the Little House books. The period was marred by sadness and strife, and the only mention of the weather is dreary and chilly.

Driven off their farm in Walnut Grove by the locust damage to their crops (more about this in Chapter 10), the Ingalls family decided to join family friends, the Masters family, to co-manage a store and hotel in Burr Oak, Iowa. They stopped near Zumbro Falls, Minnesota, on their way for a visit with relatives, the family of Peter and Eliza Ingalls. On their arrival, the already weak eight-month-old Charles Frederick "Freddie," Wilder's baby brother, fell ill. Freddie died on 27 August 1876. The heartbroken family buried him in South Troy, Minnesota, and left a short time later to complete their journey to Burr Oak.

The months of August and September were wet at the closest observing sites in Cresco, Monticello, Dubuque, and Independence, Iowa, according to FORTS data.[19] Within that wet period, 4–14 September was especially wet, with four and a half to seven and a half inches of rain among the four sites. Measurable rain fell on all but a few days of the eleven-day period at the sites. Temperatures during that week remained in the 50s to 60s for both high and low temperatures at the sites closest to Burr Oak, with little rise and fall between high and low temperatures (known as "diurnal range"), an indication that skies likely were cloudy even when it was not raining.

What little energy the Ingalls family had for paying attention to this bleak weather allowed them to see the parts that fit their dark, dampened moods.

Cold Ride from Brewster (Bouchie) School

Like many teenage girls of her era, Wilder became a schoolteacher while still a student herself. In *These Happy Golden Years,* Laura's first teaching job at the Brewster School—Wilder changed its name in the book from the Bouchie school, its real-life counter-

19. FORTS data rainfall totals 4–14 Sept. 1876: Monticello 7.44, Independence 6.00, Dubuque 5.65, Cresco 4.57 inches.

part—is plagued by cold weather, an unwelcoming boarding host, and the classroom management struggles that first-time teachers commonly faced in the one-room schoolhouse days. At the pinnacle of her eight-week job, her future husband, Almanzo, takes her home through dangerously cold temperatures. Fresh from telling Almanzo that she's only interested in his rides home and not in him, Laura is shocked when he shows up at her doorstep, right on time, on a bitterly cold Friday. She bundles up at the Brewster house, where her host is concerned for her safety. After witnessing Mrs. Brewster threaten Mr. Brewster with a knife just days earlier, Laura has no intention to stay in their house any longer than she must—the danger of a potentially mentally unstable woman with a knife must have outweighed the danger of the cold in her mind. If Almanzo plans to go home, then she would certainly go with him.

Understanding the degree of cold weather during this ride requires first correctly identifying at least the months during which Wilder taught at the Bouchie school. Following the timeline of Wilder's books literally, the teaching episode would have taken place from late December 1882 through late February 1883. The records, however, indicate that Wilder worked her first job as a teacher from mid-December 1883 through mid-February 1884, a full year later; her teaching certificate was signed in December 1883.[20]

Regardless, both winters were bitterly cold. In fact, almost every winter in the 1880s was colder than the long-term average in South Dakota and surrounding areas. At nearby Huron, South Dakota (with records from 1881–1882 through today), four of the coldest ten winters on record and five of the coldest fifteen winters are from the 1880s, even without including 1880–1881. The winter of 1882–1883 was the tenth coldest on record in Huron, and the winter of 1883–1884 was fourteenth coldest. Three of the next four winters in Huron were also among the top ten coldest, including the coldest winter on record in 1886–1887. Both winters had periods when the high temperature for the day did not rise above 0°F and low temperatures were in the -20 to -30°F range.

20. Wilder, *Pioneer Girl: The Annotated Autobiography*, p. 261; via Laura Ingalls Wilder Historic Home and Museum, Mansfield, Missouri.

In the novel, Almanzo always picks up Laura on Fridays, narrowing the search to just the Fridays from December 1883 through February 1884. The coldest of these, 4 January 1884, immediately stands out. The high temperature reported in Huron was a mere -14°F that day, with a low of -38°F. The low temperature plunged again to -35°F the next morning. No other Friday between 10 December 1883 and the end of February 1884 was nearly as cold (Friday, 8 February 1884 was the next coldest, with a high of 0°F and a low of -24°F.).

The reported temperatures never reached "40 below," as Wilder repeatedly stated in *The Long Winter* and *These Happy Golden Years*, though -38 is certainly close. The reading at Huron was so close to "40 below," however, that it very well may have been that cold in De Smet. More importantly, because mercury freezes solid at -37.89°F, the temperature could have reached -40 without anyone in De Smet being able to measure it with a mercury thermometer. Almanzo does mention that the thermometer in town froze. The thermometer in De Smet was probably mercury, whereas the observing site in Huron eventually had an alcohol thermometer paired with a mercury thermometer and recorded a temperature of -43°F on 8 January 1887. That reading of -43 remains tied for the coldest temperature on record in Huron, matched only on 12 January 1912.

The highest temperature on 4 January 1884 likely occurred somewhere around midday or early afternoon, as is common for daily temperatures during South Dakota winters. The temperature was probably already falling as the future Mr. and Mrs. Wilder began their ride and continued to fall, likely reaching its low overnight and remaining around -20 to -25°F for a good part of their trip. FORTS data around South Dakota indicates winds out of the northwest were of moderate force, likely driving northeastward with a crosswind from left to right, with wind speeds around 15 to 25 mph. The wind chill for that combination would be -48 to -55°F, which is cold enough to cause frostbite on unprotected flesh in a mere five to ten minutes. In the story, Laura and Almanzo are well bundled, but layers can only do so much. Adding in the additional 10–15 mph travel speed of the horses would only increase the wind chill factor to the range of -55 to -65°F. With all of the stops, the ride probably took an hour and a half, give or take a few minutes.

The horses start at a plucky pace, but like the cattle in *The Long Winter*, ice forms over their noses. Almanzo must stop them every couple of miles to clear their nostrils so they can breathe. Almanzo, who is surely cold himself, asks Laura questions every few minutes to make sure that she is still conscious and coherent. Laura is shockingly cold at first. Then, she begins to drift into a world where she doesn't feel the cold quite as much.

Hypothermia is sneaky. Many of us, surely, have been very cold. We shiver so hard that it hurts. Our fingertips and toes go tingly or even numb. Our gut muscles tighten. We cross the line from cold to hypothermia when the shivering stops; one enters an eerily warm place where the cold doesn't matter, becoming dreamy and removed from the harsh elements. Eventually, a hypothermic person will fall asleep as the body shuts down and, if not warmed, will die.

The cold that Laura feels on the cutter ride is not merely uncomfortable, like her bitterly cold sleeping room at the Brewster house—it is potentially deadly. Almanzo knows that Laura is hypothermic, that he is in a race against time, battling against horses that are also nearly freezing. But he wins. He gets her home to her family, where she can be warmed back to life. Safe in the arms of her family, she is given fresh clothes, blankets, a seat next to the fire, and a cup of tea. She feels like she will never get warm again (perhaps a familiar sensation after the Hard Winter), but she does. The temperatures outside begin to recover, she catches a good night of sleep, and as Ma often would say (and Laura does say), "All's well that ends well."

Every cold winter, cold snap, or even cold summer in Wilder's experience had one thing in common: they ended. Spring always arrives, bringing welcome daylight and warmth, greening and thaws, as well as its own set of dangers. No season is immune from hazardous weather.

Cold is not sneaky, at least not until the brink of hypothermia-induced death. Cold air strikes the face like a slap, especially if driven by wind. It freezes the insides of nostrils, waters the eyes, and drapes icicles into mustaches and beards. Among weather hazards, it announces itself more loudly than just about any others. It shows, too, in the number of fatalities caused by cold compared to those caused by quieter hazards like flooding and heat. Because the cold can be forecasted well in advance, and because it is readily observed by anyone who walks out the door and yelps a Midwestern "Whew, sure is brisk out here!," it is also easier to prepare for the cold, as long as someone has proper clothing and shelter for the conditions.

The protection afforded by clothing and shelter is no small matter. The Ingalls family only barely staved off hypothermia and starvation inside their home at times, with fuel sources that barely kept the cold at bay and food sources limited by their poverty or by what the trains could carry. Poverty still exists today, and many people still lack adequate heat for their homes, clothing to layer, and food to nourish them. These are the communities that still struggle most against the cold.

For those of us who are lucky enough to have warmth during winter, protecting ourselves becomes a matter of paying attention to the weather and the forecasts, and then making safe choices. It might mean that a family chooses to stay home during a bitterly cold snowstorm instead of venturing in the car to visit a friend, keeping them safe from possible travel mishaps that could expose them to the elements. It might mean that a school superintendent chooses to close schools for the day so that children aren't exposed to dangerously cold wind chills at their bus stops and on their walks to school. It might mean that a community opens a warming shelter for their unhoused people.

Knowing the hazards we face allows us to make choices about how to face them. Sometimes, the hazard masks itself or surprises us, taking some choices out of our hands.

6

Electric Storms and Cyclone Weather

It's a hot, humid summer day—the kind of day in which you break a sweat stepping out of your door even in the early morning. You reach for a sleeveless shirt or lightweight T-shirt and a pair of shorts. Sneakers would be way too confining, so you slide on flip-flops. Ready for your day, you hop in your car, which probably has air-conditioning, to run an errand to the store, where a wave of chilled air washes over you as the sliding doors open.

Errands done, you bring your goods home and put them away, again welcoming the cool air inside your house as you get ready to start on a couple of outdoor chores. Sweating through some yard work, you watch clouds billow and bloom, taking on a cauliflower shape, until you hear a distant rumble of thunder. You pack up your tools and head inside, flip the television to a local channel with your favorite on-air meteorologist, and pick up an indoor task. The green and yellow radar blotches grow orange and red kernels, and then a yellow box outlines the storm—a severe thunderstorm warning. You move the lawnmower aside to make space and pull your car into the garage, then text-message your siblings across town to make sure they know about the warning and have their kids safely inside.

Fat rain drops fall, a few at a time, then in a downpour, until they blanket the sidewalk and driveway, pooling together into an impromptu stream as they reach the road. You hear a distinct "tink" as the first hailstone falls, just a pea for now, but soon joined by the "tank" of a quarter and the "thunk" of a golf ball. Peering through the window, you cringe as your petunias and marigolds flatten under the assault. A cool gust of wind ushers the water and ice away, off to the east. As the downpour fades to sprinkles, you tiptoe outside again, looking for the biggest hailstone. You find one that you can nearly cup your hand around—at least a golf-ball size, for sure, maybe even the size of an egg—and post a picture of it to your favorite social media, remembering to send it to your sibling who isn't on social media.

Even if the conditions are identical, the way we experience the weather today is much different from how people would have during Wilder's childhood.

On the same hot, humid day, you would have woken in a sweat already, without the cooling air-conditioning to temper the muggy morning air. You would pull long trousers and a long-sleeved cotton shirt over your muslin long underwear and undershirt, affixing suspenders to your pants, with home-knitted stockings held in place with leg garters, and finally, work boots. Or you would don your muslin bloomers, at least a couple of muslin petticoats, muslin undershirt, and binding muslin corset, topping these with a long-sleeved, long-skirted calico dress with a neckline no lower than your collarbone. You would pull on home-knitted stockings and ankle boots.[1] Ready for your day, you would set out on foot or hitch the open-air buggy to run an errand to the store, which is damp and hot inside despite the open front door that lets in puffs of hot and humid air.

Errands done, you bring your goods into your equally stale and hot house, despite the open door at its front. Sweat soaking through your layers of underclothes, you are hard at work outside. You watch clouds billow and bloom, taking on a cauliflower shape, then hear a distant rumble of thunder. You hurry on with your

1. Thanks to Laura Keyes for the wardrobe details.

work until the first fat rain drops begin to turn the dirt to mud around your house.

After tending to your livestock, you reach the relative shelter of your home just before the heaviest downpour starts. You hear a distinct "tink" as the first hailstone falls, just a pea for now, but soon joined by the "tank" of a quarter and the "thunk" of a larger stone. Peering through the open front door, which you are holding ajar to let in rain-cooled air without inviting the downpour to join it, you cringe as your peas and beans flatten under the assault, hoping that enough plants survive to save at least some of the harvest for your winter food stores. A cool gust of wind ushers the water and ice away, off to the east. As soon as the downpour fades to sprinkles, you dash outside to check on the livestock, all of whom are unharmed, and to assess the damage to the garden near the house. Later, you ride your horse over to your sibling's home to check on how the family fared during the storm. Your sibling is tending to a child who was dashing home from the fields ahead of the storm. The child took a hailstone just above the eye when he was steps from the door. He has a cut and quite a bump, but ironically, the family can apply an ice compound with the melting hailstones to ease the swelling. Your families are relatively unscathed by the storm, all things considered.

Thunderstorms still exhibit the same characteristics today that they did in the late 1800s. Nothing has changed about storms themselves, but everything has changed about how we experience storms, and some things have changed about how often people experience storms in any one place, and when, and even how much rain they bring. The distant rumbles of thunder can be mistaken for a distant jet in the sky, if we aren't listening closely. The sounds of thunder and rain are harmonized by cars and traffic, calls on our cell phones, train whistles and airplanes. The view of the storms is crisscrossed by wires, punctuated by windmills, or obscured by buildings. The smell of fresh rain—known as "petrichor" when that rain falls on dry grass and dirt—may be joined by the smell of feedlots, diesel fuel, exhaust, greasy restaurants, or mowed grass, depending on where we are. In our lighter modern clothing, the warm and humid air around a storm feels less oppressive than it would have to our predecessors in long sleeves and layers.

With such visible, visceral storms, we can now and could in the past observe the signs of their approach and react to keep our families and property safe. Forecasts and warnings give us even more notice, as long as we are tuned in to receive them.

The Buggy-Ride Tornado

Stormy days on the plains often begin warm and humid. The morning smells damp, and it might even be cloudy, though any clouds will probably clear sometime in the morning and make way for the sun to warm the air. The temperature can be as warm as a mild spring or summer day, even at daybreak. Winds that may have slacked overnight pick up by mid-morning as the sun's heat creates turbulence that brings the gusts down from above our heads to ground level.

By around midday, puffy cumulus clouds build in the sky, the ones that look like piles of marshmallows and often form interesting shapes for those of us who lie on our backs and watch them roll by. As the afternoon passes, a couple of those cumulus clouds start to get taller than the rest, taller than they are wide. In technical terms, those are "towering cumulus" clouds. Meteorologists and storm chasers look for towering clouds as a sign that thunderstorm development is imminent. With the right blend of ingredients, one of those towers starts to dominate the rest, growing tens of thousands of feet tall, with a back edge that looks like cauliflower: crisp and crunchy, with an abrupt edge separating cloud from clear air. The top flattens out, with the "icing" at the top (the anvil cloud) spreading downstream ahead of the thunderstorm. It has grown up and become a cumulonimbus cloud.

In the heart of the storm, air rises quite rapidly, and it usually rotates at least slowly. Where the air is rising, the base is flat and often rain-free, with a dark shadow of rain beside it. The dark rain swath looks ominous, but the benign-looking flat base masks its danger. The most well-formed thunderstorms rotate more rapidly, and along that edge, in the rain-free area, the base of the storm lowers. This feature, called the "wall cloud," is where most tornadoes are born. A bystander watching the wall cloud will be able to see air rising and clouds swirling counterclockwise. Sometimes, a funnel cloud forms on the base of the wall cloud, pointing down toward the ground and spinning more rapidly than its parent wall cloud. In some cases, that rotation stretches all the way from the base of the thunderstorm to the ground, and a tornado is born.

Scientists still don't fully understand what makes some rotating thunderstorms produce tornadoes and not others, but a few ingredients are essential: heat and humidity, which combine to create "instability"; winds that change direction and speed with height, known as "wind shear"; and something in the atmosphere that triggers thunderstorm development, such as a front (a boundary between colder and warmer air masses, or drier and more moist air masses), a sea breeze, or even the air flowing out of another thunderstorm.

In *These Happy Golden Years,* Laura immediately notices the heat and humidity on Thursday, 28 August 1884 (not a Sunday as her book narrative would indicate). The wind blows in "hot puffs"—a warm, gusty wind, rather than a cooling and steady breeze. On most stormy days in the Great Plains and Midwest, the wind ahead of the storms will blow out of the south, southeast, or southwest. As the clouds build, Laura and Almanzo put up not only the buggy top, but a clever rain curtain that she describes in detail. Lightning shoots through and around the thunderstorm, and the young lovers sense that they're in trouble.

They are. Lightning alone is a weather killer.[2] A flash should drive anyone inside.

Then the trouble really starts as a funnel forms and reaches the ground as a tornado. Laura observes three funnels that take turns tearing at the ground. It seems, based on Wilder's description and supported by those of others, that she indeed witnessed three separate funnels that intermittently reached the ground, and not just one multiple-vortex tornado, the kind that has one large funnel but several fingers that spin around and reach for the ground as part of that single tornado. Almanzo estimates that the tornadoes are about ten miles away, approaching the couple from the north-

2. Weather Related Fatality and Injury Statistics, National Weather Service, accessed 7 Feb. 2024, https://www.weather.gov/hazstat/.

west and moving to the southeast as they race toward the northeast ahead of them. (His memory was probably quite accurate; though Wilder tended to exaggerate distances in her books, Almanzo had a clearer sense of distance.) They are gambling that they will clear out of the twisters' path in time.

Given their options in the situation, the choice they make is a good one. The safest way to escape the path of a tornado is to figure out what direction it's moving in and then take a hard right and move away. For example, if a tornado is moving from west to east, one should turn south out of its path. As long as one can keep the tornado in sight and stay clear of it, and as long as one's road doesn't run out, this is the best bet. Laura and Almanzo take a left instead of a right, the more dangerous option because it puts them in the path of the hail that comes with a tornadic thunderstorm. But, it still gets them out of the path of the tornado, which passes to their west. They feel a cool breeze from the rain—and hail-cooled winds—the outflow of the storm. They safely return to the Ingalls homestead.

The closest tornado to De Smet on that day struck near Howard, some twenty-two miles from De Smet at its closest point.[3] The Howard tornado was famously photographed by F. N. Robinson, whose photograph is often misrepresented as the "first" known surviving picture of a tornado (it is actually the second such picture; the earliest photograph is from 26 April of the same year, near Westphalia in Anderson County, Kansas).[4] The three-funneled appearance of Robinson's photograph has been criticized by meteorologists through the years as potentially doctored, because its appearance is not common for tornadoes, but given Wilder's account of the three-funneled tornado as well as one additional surviving

3. John T. Snow, "Early Tornado Photographs," *Bulletin of the American Meteorological Society* 65, no. 4 (1984): 360–64, https://doi.org/10.1175/1520-0477(1984)065<0360:ETP>2.0.CO;2.

4. NOAA Digital Collections, https://photolib.noaa.gov/Collections/National-Weather-Service/Meteorological-Monsters/Tornadoes/emodule/643/eitem/2777. Even meteorologists get it wrong. For example, *see* "Oldest Known Photo of a Tornado—August 28, 1884," NWS, accessed 7 Feb. 2024, https://www.weather.gov/ffc/oldtorn.

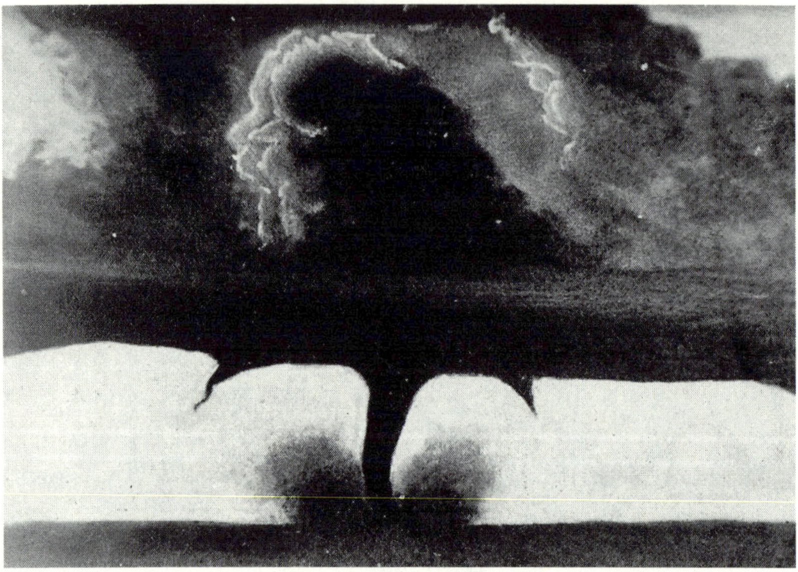

This photo by F. N. Robinson depicts a tornado about twenty-two miles south or southwest of Howard, Dakota Territory, on 28 August 1884. Another tornado that day passed very close to Howard and may have been the one seen by Laura Ingalls and Almanzo Wilder on their buggy ride south of De Smet. Whether—or how much—this photograph was retouched is a matter of debate among scholars. South Dakota State Historical Society

sketch that also depicted three funnels with intermittent contact with the ground, it seems that the photograph may be accurate, after all.[5] Later analysis has suggested that while the parent thunderstorm cloud may have been retouched—just as photographers today enhance the contrast and color of their storm photography—the tornado and funnels in the Robinson photo were not.[6]

Following her depiction of their harrowing buggy ride, Wilder included firsthand details from tornado damage that she, Almanzo, and Pa likely did not witness. Still, the stories Wilder told match other historical accounts.[7] As Wilder describes it, two boys on mules

5. Thomas Grazulis, *Significant Tornadoes, 1680–1991: A Chronology and Analysis of Events* (Environmental Films, 1993), p. 632.

6. Snow, "Early Tornado Photographs."

7. Grazulis, *Significant Tornadoes*, pp. 630–32.

are struck by a tornado during the same outbreak, and one of those boys is killed. The boys live on a homestead that was destroyed by the tornado just northeast of Forestburg, twenty-eight miles due west of Howard, right along the James River near the beginning of the tornado track that crossed Sanborn and Hanson Counties.[8] In the novel, Wilder recalls Pa and Almanzo traveling on horseback to see if any tornado victims need assistance. The actual tornado path from the Forestburg tornado was, at its closest, some thirty-five to forty miles from De Smet as the crow flies, too far for an afternoon trip via horse. Perhaps Pa and Almanzo went to help on another day, or perhaps they helped with the Howard tornado aftermath, or perhaps they didn't go at all, but rather heard the stories via their local newspapers or by word of mouth. Based on historical analysis of newspaper accounts of the damage to the boys' farm, researchers assigned a rating of F4 to that particular tornado.[9]

Tornadoes have been known to move papers dozens or even hundreds of miles downwind and to loft relatively light debris high into the atmosphere.[10] Wilder's tall tale of a door lofted by the tornado that fell directly back down to the homestead some hours after the tornado had passed stretches credulity, however. It is possible that the tornado lofted the door well into the atmosphere and that it didn't fall back down until after the storm had passed. If there was truth to Wilder's door story, it did not gain entrance into the later documentation by tornado researchers to support the F4 damage rating.

Overall, records point to at least five significant tornadoes (F2 or stronger) in eastern South Dakota on 28 August 1884.[11] The tornado closest to Howard is the one in the justly famous picture and sketch,

8. Snow, "Early Tornado Photographs," Figure 2.

9. Grazulis, *Significant Tornadoes*, p. 631.

10. John A. Knox, Jared A. Rackley, Alan W. Black, Vittorio A. Gensini, Michael Butler, Corey Dunn, Taylor Gallo, Melyssa R. Hunter, Lauren Lindsey, Minh Phan, Robert Scroggs, and Synne Brustad, "Tornado Debris Characteristics and Trajectories During the 27 April 2011 Super Outbreak as Determined Using Social Media Data," *Bulletin of the American Meteorological Society* 94, no. 9 (2013): 1371–80, https://doi.org/10.1175/BAMS-D-12-00036.1.

11. Grazulis, *Significant Tornadoes*, pp. 630–31; Snow, "Early Tornado Photographs."

and it is most likely the one spotted by Laura and Almanzo. It was rated F2 based on the damage to a farmstead and barn, and it killed thirty cattle. The fifth tornado, not seen on the map, occurred very near Sioux Falls, southeast of the maps edge. These tornadoes are known to have killed seven people.

Though De Smet was not directly affected that day, it had its own brush with storms a month earlier in July 1884. On 21 July, an F4 tornado struck Dell Rapids (fifty-seven miles southeast of De Smet, or nineteen miles north-northwest of Sioux Falls) and moved southeastward into Minnesota.[12] On 25 July, strong storms swept directly through De Smet. Eliza Jane Wilder, Almanzo's sister, recalled:

> One day I was caught in a storm about 3/4 of a mile from home. The wind blew so hard as to crush the Catholic Church and several buildings were moved from their foundations. The town hall among the rest. Hail and rain came in torrents and with such force that my arms were black and blue after. Gardens were destroyed, wheat and other grains seriously injured. Some fields were never cut. But I faced the storm for over one half mile. My sister's [Laura Wilder Howard's] children were home alone and I feared they might be seriously frightened else I should have stayed down on the ground, until it was over.[13]

The *De Smet Leader* on 26 July confirmed: "A terrible storm struck De Smet last night, and considerable damage was done. The unfinished Catholic church was blown down. Loss $800. Insured for $500. The band hall was moved about six feet and badly racked. W.S. Lawman's residence was moved from its foundation. The storm consisted of rain and wind, with a little hail, but not enough to do any damage here. The hardest of the storm is thought to have passed east of us, but no reports have yet been received."[14]

The same issue also updated "The Great Hail Storm" of "last Monday, which extended through Dakota and across the State of

12. Ibid., p. 630.

13. William T. Anderson, *A Wilder in the West: The Story of Eliza Jane Wilder* (Anderson Publications, 1991), p 20.

14. *De Smet Leader*, 26 July 1884, p. 2, https://www.loc.gov/resource/sn00065129/1884-07-26/ed-1/?sp=2&st=image&r=0.076,0.01,0.49,0.232,0.

Minnesota" and "did a fearful work among the growing crops and in some of the villages in some of its tracks." The swath, "six to ten miles wide," clipped the southeast corner of Kingsbury County, "about a dozen miles" from De Smet. The paper noted significant crop losses on numerous farms, along with damage to buildings in the village of Carthage and to farm buildings.[15] Superintendent George Williams, who tested and granted Laura Ingalls her teaching certificate in a scene at the end of *Little Town on the Prairie*, lost 145 acres of "fine crops" and "two horses which were killed by lightning."[16]

It seems that 1884 was quite a stormy summer in eastern South Dakota!

The unintended buggy storm chase was not the Wilders' only tornado encounter. A tornado clipped the Wilders' Rocky Ridge farm on 12 April 1945. The tornado, rated F3, followed a thirty-two-mile path that ended near Mansfield, Missouri. In a letter to her sister Carrie on 21 May, Wilder wrote:

> Well! The big oak trees in front and west of the house are all cleared away. A cyclone did it. It uprooted them, just tore them up by the roots. It tore out a big walnut tree by the barn and split the elm by the iris bed. The great, old oak close in front of the house was broken in pieces and the trunk split the whole length. Some shingles were torn off the house and the big south window in the front room was blown out, but we were lucky the house didn't go. A good many houses did, and people were hurt. It was two weeks before we could get the big trees out of our driveway so we could get out. Telephone and electric wires were down two weeks.[17]

The Dakotas and Missouri still see tornadoes every year, threatening lives and livelihoods. Tornado watches and warnings give advance notice so that people can take shelter. Sometimes, if one is attuned to the weather like Laura and Almanzo were at the outset of

15. Ibid.

16. Laura Ingalls Wilder, *Little Town on the Prairie* (1941; New York: Harper Trophy, 1971), pp. 303–6; *De Smet Leader*, 26 July 1884, p 2.

17. William T. Anderson, ed., *The Selected Letters of Laura Ingalls Wilder* (New York: Harper Perennial, 2017), p. 273.

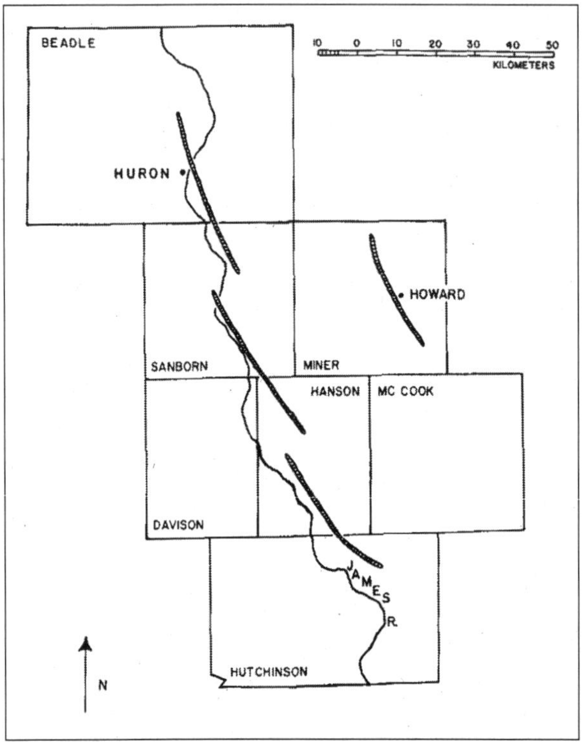

Numerous tornadoes struck during the outbreak on 28 August 1884. The closest tornado to De Smet is the one that passed near Howard and likely is the one that Laura Ingalls and Almanzo Wilder saw on their buggy ride. Recreated from a map originally published by John T. Snow, "Early Tornado Photographs," in the Bulletin of the American Meteorological Society, *April 1984.*

their buggy ride, one may even notice the signs of thunderstorms to come. This puts one on higher alert, more vigilant than one would be on an average sunny day, anxiously watching for those storms to build.

Ball of Fire

Sometimes, the atmosphere's hazards arrive much more unexpectedly. Embedded within Wilder's description of the hardest winter in *On the Banks of Plum Creek*, Laura, Mary, Carrie, and Ma observe a phenomenon like ball lightning while at home during a blizzard:

> A ball of fire was rolling down the stovepipe. It was bigger than Ma's big ball of yarn. It rolled across the stove and dropped to the floor as Ma sprang up. She snatched her skirts and stamped on it. But it seemed to jump through her foot, and it rolled to the knitting she had dropped.

Ma tried to brush it into the ashcan. It ran in front of her knitting needles, but it followed the needles back. Another ball of fire had rolled down the stovepipe, and another. They rolled across the floor after the knitting needles and did not burn the floor.

"My goodness!" Ma said.

While they watched those balls of fire rolling, suddenly there were only two. Then there were none. No one had seen where they went.[18]

Ball lightning remains an atmospheric enigma, rarely observed, poorly documented, and physically not well understood. Virtually all observations and descriptions of ball lightning are historical and anecdotal, with little or no photographic or video evidence. Scientists have yet to reliably create ball lightning in laboratory settings. They have, however, determined three basic characteristics of ball lightning: a duration of some seconds, a floating motion, and sudden disappearance. The few research studies on ball lightning have focused on the theoretical physics and chemistry principles that could explain the phenomenon, some of which were sparked by reported ball lightning sightings. One study suggests the orbs could be composed of photons trapped in an air bubble of their own making, refracting light back into itself.[19] Another research team hypothesizes that ball lightning is caused by oxidization of silicate nanoparticles (that is, tiny bits of sand-like fragments) as lightning strikes the ground, which eject and link in the air and then release their stored energy as heat and light.[20]

18. Laura Ingalls Wilder, *On the Banks of Plum Creek* (1937; New York: Harper Trophy, 1971), p. 310.

19. V. P. Torchigin, "Physics of a Ball Lightning in a Form of a Bubble of Light," *Optik* 188 (July 2019): 294–301, https://doi.org/10.1016/j.ijleo.2019.05.009. A supporting article to help break down the harder science is available from Paul Scott Anderson, "Has the Ball Lightning Mystery been Solved?," *EarthSky*, 14 June 2019, https://earthsky.org/earth/ball-lightning-lightning-atmosphere-earth-optik.

20. John Abrahamson and James Dinniss, "Ball Lightning Caused by Oxidation of Nanoparticle Networks from Normal Lightning Strikes on Soil," *Nature* 403 (February 2000): 519–21, https://doi.org/10.1038/35000525 (also referenced in Anderson, "Ball Lightning Mystery").

At the very least, Wilder's description of her family's ball lightning encounter aligns with descriptions from other sources. Its size, "bigger than Ma's big ball of yarn," is consistent with other reports of it being the size of a grapefruit, or "the size of a child's head."[21] Most reports indicate the ball lightning lasting under a minute, with a rainbow of reported colors.[22] Luminescent balls have been reported both indoors and outdoors. If indoors, they are described as either entering through a door, window, or chimney or as materializing inside the building. The phenomenon has been reported both with and without more traditional lightning in the area, and it may end with a bang or with no sound at all.

Scientists often look for other plausible explanations when examining a potential rare event. In this case, it is difficult to imagine other causes for elusive balls of light indoors. A round, dried plant-like object on fire, such as a tumbleweed, would have a similar shape, but it also would have burned to the touch, given off smoke, and likely would have lit other objects (like Ma's skirt) on fire as it rolled across the floor. A light phenomenon, such as a refraction or reflection of some kind, would not have rolled across the floor, nor would it have been perceived as a three-dimensional object. We will never know if the Ingalls girls truly did witness rare ball lightning, but for now, it's hard to imagine what else it might have been.

The Poor Man's Ice

While almost any thunderstorm can produce hail, it is the supercell thunderstorms—the tall, rotating storms—that produce the largest

21. Anderson, "Ball Lightning Mystery"; "Periodically I Hear Stories about Ball Lightning. Does this Phenomenon Really Exist? Could a Ball of Plasma Remain Stable for Several Seconds, as Some Researchers Have Claimed?," *Scientific American*, 18 July 1997, https://www.scientificamerican.com/article/periodically-i-hear-stori/.

22. Ibid.; *see also* Eric Canan, "Ball lightning is so Strange it Might Just Come from Another Dimension," *New Scientist*, 21 Oct. 2020, https://www.newscientist.com/article/mg24833053-000-ball-lightning-is-so-strange-it-might-just-come-from-another-dimension/; Anderson, "Ball Lightning Mystery; Christina Nunez, "Ball Lightning: Weird, Mysterious, Perplexing, and Deadly," *National Geographic*, 5 Mar. 2019, https://www.nationalgeographic.com/environment/article/ball-lightning.

hail. The strong updrafts loft tiny particles of dust, dirt, and pollen through the rain, then above the freezing level high in the sky, where the coating of rain turns to ice. The tiny particles become hail embryos, the seeds around which hail forms. The nascent hailstone reaches the top of the updraft, and losing the upward thrust of air, begins to fall. If the updraft is strong, it will pick up the baby hailstone again, lifting it back through the rain and above the freezing level, adding a layer of ice like a coating of wax on a homemade candlestick. The process can repeat a few times or dozens of times, the hailstone growing larger with each trip through the rain and above the freezing level. Professional observers and amateur storm spotters alike estimate hail size by comparing the biggest hailstones to coins, sports balls, and some fruits and other objects. Meteorologists prefer measurements to estimates, but measuring hail is challenging. The largest stones often fall unseen or break when they hit the ground or melt before the storm has passed and it is safe for observers to retrieve them. Also, hail is not perfectly round, but usually oblong—longer on one axis than the others—and it is sometimes spiky. Meteorologists want to know the very longest diameter of the very biggest stone in order to warn people about the very worst of the risk.

When the hail becomes too heavy for the updraft to loft it again, the thunderstorm births the full-grown hail. It falls to the ground, perhaps straight down or perhaps driven by the strong winds. The largest hailstones are closest to the updraft area of the thunderstorm, while smaller ones are flung farther downwind by the stronger winds in the middle of the thunderstorm. As a hailstorm moves over a fixed point, rain begins, with drops becoming fatter. A few small hailstones join the mix, then the larger ones. If the updraft of the storm passes directly overhead, the stones may stop for a moment, or there may be just a few stones dropping. Another round of rain and hail follows, with the sizes and numbers of the hailstones decreasing as the thunderstorm moves on.

A hailstone falls at its terminal velocity, which takes into account the size, weight, and often irregular shape of the hailstone. A golf-ball-sized stone weighs around 0.8 ounces and falls at about 60 mph, while a baseball-sized stone weighs about 8 to 12 ounces and falls at about 80 mph. Very large hailstones, four inches in diameter and larger, can weigh over a pound and fall at over 100 mph. The largest hailstone on record in the United States, a giant 18.62 inches in diameter and weighing almost two pounds, fell in Vivian, South Dakota, on 30 June 2010. The force of a falling hailstone of golf-ball size or larger can bruise flesh, dent cars, and break lights, siding, gutters, and shingles. Baseball-sized hailstones can leave welts and, if they strike an unfortunate head, cause concussions, as well as break through windows, windshields, and siding.

Wind only adds force to the hailstones. Even dime-sized hail driven by 70 mph winds can shred mature corn and soybean plants. Golf-ball-sized hail driven by 70 mph winds can break windows in homes, loosen shingles, dent siding and soffit, and shatter windshields. Baseball-sized hail driven by the same 70 mph winds can debark trees, wreck homes, total cars, and kill unsheltered livestock and humans.

The NWS considers any hailstone with a diameter of 1 inch (quarter-sized) or larger to be "severe," warranting a severe thunderstorm warning. Hail that is 2 inches ("hen egg"-sized) or greater is "significant." Hail that is 2.75 inches (baseball-sized) or greater is considered "destructive" and will trigger an extra level of alerts through cell phones.

Laura's neighbor in *The First Four Years* violates common safety rules when he goes outside during a hailstorm to collect hail. Unsurprisingly, a hailstone knocks him unconscious. Wilder writes that the hail is "as large as hens' eggs." The damage she reports includes broken screens and windows, trees stripped of branches and leaves, and, of course, the flattened wheat crop. Wilder includes a bit of ironic folk wisdom in the novel: "The rich have their ice in the summer but the poor get theirs in the winter."[23] Wryly, Laura observes that the poor got their ice this summer.

The hail devastates the Wilders' nearly mature wheat crop in their first year of marriage, sinking them into debt. Though Almanzo suggests they make ice cream with the ice from the hailstones, the storm is a crushing blow to their farming enterprise.

23. Laura Ingalls Wilder, *The First Four Years* (1971; New York: Harper Trophy, 1972), p. 4.

Wilder's hailstorm fell long before the National Weather Service provided severe thunderstorm warnings for such hail. Even so, nothing can be done to protect a field of crops from a swath of hailstones, then or now. Hail damage has been covered by crop insurance since 1880, and modern farmers can receive insurance payout benefits if their crops are damaged, allowing them to maintain operations from year to year despite the bad luck of being in a hail swath.[24] Insured crop losses due to hail damage can be roughly $20 million to $300 million per year in the United States.[25] Home and auto insurance also provide property damage protection, allowing residents of hail-prone areas to manage the costs of replacing roofs, siding, fences, and cars when hail strikes.

Manchester's Fate

Manchester, roughly seven miles west of De Smet, was the next stop west along the railroad tracks. It was a small town, not all that different from De Smet when it was founded, with a few hundred residents and a few businesses: a store, a livery barn, a depot, a restaurant, schools, churches, and gas stations and service stations once automobiles arrived

In *These Happy Golden Years*, Laura spends several weeks of a summer keeping dressmaker Mrs. McKee company on her homestead near Manchester. Much later, Grace Ingalls met farmer Nathan Dow, who had a farmstead near Manchester, while teaching there. They married in 1902 and lived on Dow's farm for several years, until health and finances led them to rent out the farm. After spending time living in De Smet and even briefly in Oregon, Grace and Nathan Dow settled in a rented home in Manchester, where they lived out their remaining lives.

By the early 2000s, the population of Manchester had dwindled to just forty residents, most in farmsteads outlying the town itself, with just six souls in the town's footprint. Most of the businesses in Manchester had folded, with the town comprising about a dozen

24. National Crop Insurance Services, "What Is Crop Insurance?," updated May 2025, https://cropinsuranceinamerica.org/what-is-crop-insurance/.

25. On payouts, *see* https://www.statista.com/statistics/1015612/crop-hail-damage-usa/; https://www.rma.usda.gov/SummaryOfBusiness/CauseOfLoss.

buildings, including two active businesses and a post office, and three or four remaining families, along with a well pump in the middle of town.

On 24 June 2003, the largest tornado outbreak in South Dakota history struck the state. Sixty-seven tornadoes raked farmlands and towns east of the Missouri River. From the strongest of these supercells, a tornado stretched from the base of the storm to the ground, stirring up dirt in open fields. This first tornado, weak and brief, was quickly followed by a second, stouter tornado. Ballooning to a half mile to one mile in width, the wedge tornado careened through farmland and homesteads, then moved from south-southwest to north-northeast, crossing Highway 14 at precisely the location of the town of Manchester. An F4 tornado would destroy all but the sturdiest of structures, and the aged homes and businesses in Manchester proved no match. What little remained of the town was reduced to matchsticks and fragments and blown away in the winds of the tornado.

The remaining residents of Manchester opted not to rebuild, and in 2005, South Dakota officially declared it a ghost town. Today, all that remains is the well pump and a memorial plaque, built in part by one of the former residents of a nearby homestead. Surrounding the marker in the former town center are the bare outlines of the buildings and a grid of former gravel roads, nearly overgrown as weeds and grasses encroach.

As Ma would say, there is no great loss without some small gain, and though the world lost the town of Manchester, it gained a treasure trove of scientific knowledge from the Manchester tornado. Storm researcher Tim Samaras, his research partners, and a National Geographic Society film crew targeted the Manchester storm on 24 June.[26] Samaras, a self-taught engineer whose formal education ended in high school, had combined his lifelong passions for tornadoes and tinkering into a fifty-pound cone-shaped instrument probe tossed into the back of his Dodge Caravan. Dubbed "turtles," the probes contained instruments to measure atmospheric pres-

26. Biographical material about Tim Samaras is drawn from Brantley Hargrove, *The Man Who Caught the Storm* (New York: Simon & Schuster, 2018).

sure. While meteorologists have had success in deploying weather radar and measuring equipment around tornadoes, Tim's goal was to place the turtle into a tornado. And on 24 June, it worked.

Early attempts to place instruments into the paths of tornadoes began in the 1980s. A team of severe weather researchers with the National Severe Storms Laboratory designed a bulky instrument package with sensors for temperature, moisture, wind, and pressure. The Totable Tornado Observatory, or TOTO, would look familiar to fans of the movie *Twister*: the movie's Dorothy instrument was copied from the NSSL experiment.[27] Both TOTO and Dorothy reside today on the first floor of the National Weather Center in Norman, Oklahoma, greeting visitors to Norman's weather mecca.

TOTO was backed by some of the greatest minds in tornado research, but it failed its mission; it was never deployed successfully into the path of a tornado. Its two-hundred-pound weight and bulky frame made its deployment clumsy and slow, when movements in front of a tornado (if someone is brave and compelled to make them) must be nimble and agile. Had it been hit by a tornado, it likely would not have survived anyway; later tests showed that the instrument, as with Dorothy in the movie *Twister*, was too light to stay upright and on the ground in anything but a weak tornado.

Samaras was charged with designing an instrument that one person could handle. It had to be agile enough for rapid deployment in the path of a tornado and facilitate a rapid exit out of harm's way, but sturdy enough to withstand the winds of a violent tornado, the higher-intensity EF4 and EF5 winds that carry away entire buildings. Samaras's turtle not only resisted being lifted, it actually compressed farther into the ground under the strain of tornado winds. As one might expect, it is difficult to both accurately forecast the exact path of an ongoing tornado and navigate into the path and out again safely; the margin of safety is paper-thin and shifts constantly. Samaras and his team had several near-misses and a sideswipe on a turtle before their chase on 24 June 2003.

Samaras watched the first tornado develop and dissipate, then

27. NOAA/NWS/Storm Prediction Center FAQ, "The TOTO Home Page," accessed 16 Feb. 2024, https://www.spc.noaa.gov/faq/tornado/toto.htm.

the much stronger second tornado descend soon after it. Following the tornado's diagonal movements along the perfectly square-gridded road, often on soggy and slippery gravel, his two-car team zigzagged, dropping one turtle along the way, then ducked into a gravel intersection like a competitor on an obstacle course. Dropping a series of turtles and camera probes on the side of the northbound gravel road, the team zipped out of harm's way in the nick of time. The camera probes contained video cameras to record a 360-degree view of the approaching wedge tornado.

The tornado made a direct hit on two turtles and a camera probe. One turtle was laid about a mile west of Manchester, and the other turtle and camera probe were about two miles north-northeast of town. The tornado connected the dots from turtle to town to turtle.

The groundbreaking data collected by Samaras included the steepest pressure drop ever recorded as the tornado passed directly over the turtle and its sensors.[28] The 100-millibar drop is the difference between an average rainy or snowy low-pressure system and the fiercest hurricanes on record, and it demonstrated to researchers how quickly the pressure changes across the width of a tornado. Combined with National Geographic's professional camera crew filming both thc instruments' deployment and the tornado itself, the Manchester tornado became one of the most well-documented in history. The National Weather Service rated the tornado an F4 on the Fujita scale, among the strongest ever recorded in the state.[29]

Samaras continued to develop and deploy instruments in and around tornadoes, seeking more data, but his life's work became his fated legacy. On 30 May 2013, Samaras and his small team (including his adult son, Paul) were swallowed by a powerful and fast-moving tornado near El Reno, Oklahoma, killing all three in their car. Three turtles had been staged in the trunk of the car, undeployed. At the time of this book's publication, Samaras and his team are the only researchers to lose their lives while undertaking the dangerous task of taking direct tornado measurements. He devoted his life to

28. National Weather Service Sioux Falls, South Dakota, "Record Pressure Drop Recorded with Manchester Tornado—24 June 2003," accessed 16 Feb. 2024, https://www.weather.gov/fsd/20030624-tornadosamaras.

29. NCEI Storm Data.

observing and studying tornadoes and he died in dedication to that work.

Hurricanes

Though the Ingalls family never lived close enough to the tropics to be impacted by a hurricane, the Wilders spent a year of their adulthood in Florida. In 1891–1892, Laura and Almanzo Wilder tested their fortunes in the Florida panhandle, near the town of Westville. There were no impacts by any hurricanes or tropical storms during the months the Wilders lived there. Nonetheless, neither the weather nor the culture agreed with them, and Mrs. Wilder found both particularly displeasing.[30]

No other technology has helped with hurricane tracking and forecasting as much as satellites. The earliest pictures in the 1970s shed a dim but vital light on the structure of hurricanes and how it compares to their intensity, and the satellites provided valuable eyes in the sky to track hurricanes across areas with no other record keepers available to relay reports. Before the satellite era, hurricane tracks were reconstructed based on ship observations and records, along with records and stories of the destruction as hurricanes made landfall. Though scientists have been able to complete a remarkable number of reconstructions using these observations, in a scientific game of "connect the dots" it is almost certain that entire hurricanes went unreported before satellites could see them.

Westville has been impacted by numerous hurricanes and tropical storms in its history, most recently by the landfall of Hurricane Michael, a Category 5, in October 2018.[31] That hurricane devastated coastal areas near Tyndall Air Force Base and Panama City, Florida, and caused widespread tree loss inland. Weather instruments recorded a 102-mph wind gust as far inland as Marianna, Florida, just thirty-eight miles east of Westville. The center of a tropical storm

30. Laura Ingalls Wilder, "Autobiographical Sketch for 'The Junior Book of Authors,'" in *LIW: The Little House Books*, vol. 2, p. 799; and Donald Zochert, *Laura: The Life of Laura Ingalls Wilder* (New York: Avon Books, 1976), p. 194.

31. Past hurricane landfalls are available from NOAA/NWS/National Hurricane Center, "Data Archives," accessed 16 Feb. 2024, https://www.nhc.noaa.gov/data/.

or hurricane track has directly crossed the small town of Westville three times in archived history: November 1861 (tropical storm), September 1903 (Category 1 hurricane), and November 1904 (tropical storm).

Ten known tropical storms or hurricanes formed in the Atlantic Ocean basin in 1891, the summer the Wilders were there. The first one of the year made landfall in early July just west of Galveston, Texas, as a Category 1 hurricane. Its strongest punch may have been thrown into Louisiana, where tornadoes associated with the hurricane damaged buildings and a prison in and near Baton Rouge. The third hurricane of the year was the strongest, making landfall in Martinique, an island in the Caribbean Sea, as a Category 3, killing over seven hundred people there. In its final gasps, now as a Category 1, it crossed the Florida peninsula from near Homestead to near Naples. It dissipated shortly after entering the Gulf of Mexico, before it had a chance to impact the panhandle.

In 1892, nine known hurricanes or tropical storms formed in the Atlantic Ocean basin. The closest to the Florida panhandle made landfall in mid-September near the Louisiana Mississippi River basin, where it likely brought at least rain to the panhandle. The Wilders had departed the region about a month earlier. Had they stayed through 1894, they would have experienced a hurricane when a Category 3 made landfall near Panama City, Florida, on 8 October. The fifth of seven known hurricanes of the season, it tracked from there through the southeast and toward Massachusetts into Canada. The only known fatalities with this hurricane occurred much farther north, as wind damaged a tenement in New York City and caused a roof collapse on a home in New Jersey.

Laura and Almanzo Wilder's daughter, Rose Wilder Lane, found herself more exposed to tropical threats. Lane kept a home in Danbury, Connecticut, from 1938 until her death in 1968. While Connecticut is not as prone as other stretches of the Atlantic coast to hurricane strikes, it can be impacted from time to time. Lane also had a home in Harlingen, Texas, in the southern tip of the state, from 1965 until her death in 1968, an area at higher risk from tropical storms and hurricanes.

On 26 September 1938, Wilder wrote to Lane, "Your letter dated the 22nd came this morning. It was evidently written before the storm was at its worst and we are anxious to hear from you again. It seems bad enough the way you wrote and we are hoping the bad storm passed you by and that your electricity being off and wires down is only your part in that terrible storm. But you did say 'the clouds look threatening.'"[32]

The storm Wilder referenced was the Great New England Hurricane of 1938, which made landfall on 21 September, first on Long Island, New York, and again between Bridgeport and New Haven, Connecticut. The hurricane was a Category 3, with winds estimated at 115 mph, and it carried a twelve- to fourteen-foot storm surge that caused extensive coastal flooding, damage, and deaths. Nicknamed the "Long Island Express," the hurricane crashed into Long Island and New England with virtually no warning, due to a combination of weather and human factors.[33]

Lane was living in Danbury at the time, nineteen miles northwest of Bridgeport and near the New York state line. While her home would have been far enough inland to escape the coastal flooding, there is no doubt that she experienced hurricane-force winds and extremely heavy rain for hours as the center of hurricane passed just miles to her east. The extent of damage to her home is not documented, but Wilder likely surmised correctly that the area experienced power outages and downed lines, with the devastating loss of over two billion trees across New York and New England and with around 20,000 miles of electrical power and telephone lines downed across the region.[34]

It is a testament to the fortitude of the United States Postal Ser-

32. LIW to RWL letter, 26 Sept. 1938.

33. Lourdes B. Aviles, *Taken by Storm, 1938: A Social and Meteorological History of the Great New England Hurricane* (Boston: American Meteorological Society, 2013) and R. A. Scotti, *Sudden Sea: The Great Hurricane of 1938* (New York: Little, Brown and Co., 2003). *See also* National Weather Service, "The Great New England Hurricane of 1938," https://www.weather.gov/okx/1938HurricaneHome.

34. Scott A. Mandia, "The Long Island Express: The Great Hurricane of 1938: Damage Caused by the Storm," updated 9 Feb. 2018, https://www.sunysuffolk.edu/explore-academics/faculty-and-staff/faculty-websites/scott-mandia/38hurricane/damage_caused.html.

vice that a letter dated 22 September would arrive in Mansfield by the twenty-sixth, given the chaos and destruction across New England. If Lane wrote the letter before the storm arrived, then it was probably written and mailed on 20 September, as it is unlikely that she would have felt the effects or been aware of a hurricane more than a day in advance.

Years later, Lane found herself in the path of Hurricane Beulah as it struck southern Texas on 20 September 1967, near her home in Harlingen. Lane wrote later to her friend Jasper Crane that it was a "unique and marvelous experience which I wouldn't have missed." In an interview, Carol Griffen Mayfield, who lived next door to Lane in Harlingen as a child, noted Lane's fondess for sharing tales of the 1938 Great New England Hurricane, explaining, "We heard countless stories about that hurricane so when Beulah arrived, we were quite determined to 'weather the storm.' It came at night and was very scary and thrilling at the same time."[35] Lane's home escaped damage from Hurricane Beulah, but it seems that she enjoyed the chaos and power of the storm around her.

Landfalling hurricanes often give birth to tornadoes as they move ashore. Though they can spin up in any sector, the side of the storm to the right of thc eye, or center, is where tornadoes are most likely to develop. That side is where winds come onshore and where they create counterclockwise curl, aiding the kind of spin in the atmosphere that favors tornadoes. The abundant tropical warmth and moisture provides enough instability for tornadoes, while the howling winds near and above the ground often provide just the right kind of wind shear to create spin. A hurricane does not have to be strong to be a tornado producer, just as it does not have to be strong to be a flood maker; it is more important to have an optimal blend of ingredients than to have the strongest of the winds or deepest central pressure in the hurricane itself.

Hurricanes occur over warmer oceans, as warm water is to hurricanes what coal is to a steam engine: fuel. Because of the physics of hurricanes, that extra fuel leads to extra intensity. For now, the

35. Susan Albert, "Rose Wilder Lane in Texas," accessed 16 Feb. 2024, https://susanalbert.typepad.com/lifescapes/2014/09/rose-wilder-lane-in-texas.html.

effects of the warmer oceans on hurricanes still have scientists in heated debates, the most vociferous of which center on how much the warming will cause hurricanes to intensify (not a question of if, but rather of how much) and whether there will be more hurricanes, fewer, or the same number.

Some aspects are less debated. The storm surge—the wall of sea water pulled up by a hurricane and dragged ashore—occurs on top of rising sea levels. To return to an earlier example, the child in the bathtub who pushes waves on top of water in a half-full tub will flood the bathroom much less than one who pushes waves on top of water in a full tub. As the sea level rises, the tub fills, and the waves can reach higher, farther inland, flooding areas never flooded by surges in the past. Also, the moisture that has been added to the atmosphere because it is warmer adds water into the hurricane's system, bumping up the already torrential rainfall that hurricanes bring. It may remain difficult to define exactly how much of a hurricane's rainfall was added on by the warming of climate change, but it is an undercurrent, just like steroids were an undercurrent to home run records in a certain era of baseball.

Even without the added fuel, hurricanes have the power to do more damage on coasts simply because of all that has been built there. People and businesses in the path have days of advance notice to prepare. They can take measures to protect their homes and can evacuate to protect their lives. But sometimes, even the best protection is not enough in the face of disaster.

While many tornadoes are highly visible and well tracked, and hurricanes show themselves in vibrant satellite images days before landfall, sometimes even these most visible and attention-grabbing storms sneak up on us. A tornado can be obscured by rain or darkness, or it may be stealthily out of sight of the hard-working meteorologists watching for it on radar. A hurricane may make just a slight shift in track at a narrow angle to the coastline and create dramatic impacts, or it may intensify or weaken rapidly just before landfall.

The character of a tornado or hailstorm or hurricane has not changed from Wilder's years to ours. The seasons of these storms may be shifting slightly, or the locations, or the intensity. But tornadoes and hail still threaten the Dakota prairies, and hurricanes still threaten the coasts of Florida, Texas, and Connecticut. Seasoned residents often have trusted sources of information to alert them when danger threatens. Travelers may rely on those seasoned residents, or trusted but not-so-local sources. They might not have a favorite local TV meteorologist, for example, but they will listen raptly to The Weather Channel and trust words of warning from a well-known coastal post.

Every cell phone is equipped to receive Wireless Emergency Alerts for the biggest hazards—tornado warnings, the highest-end severe thunderstorm warnings (with baseball-sized or larger hail, or 80-mph winds or stronger), hurricane warnings and evacuation notices, and high-end flash flood warnings. The alerts only buzz the phones that are included in a National Weather Service warning, without needing to know ahead of time what county you may be visiting (though it's still helpful).

Of course, receiving a warning or alert, or hearing from a trusted source, is only part of the path to safety. Staying safe also means knowing what to do when that phone buzzes and you're in a warning: where safe shelter is, how to prepare, and even when it's safe to come back out of shelter again. Even with their hidden dangers and mysteries, tornadoes, thunderstorms, and hurricanes give more visual and auditory notice of their appearance than some other hazards do, ones that approach stealthily and hide their danger until it is too late.

7

Freshets and Floods

In the middle of the United States, floods usually come from melting snow or heavy rain. As snow melts and ice breaks up in the spring, the water filters into streams and creeks, then spills into smaller rivers, which converge into the large rivers and flow to the ocean. Swelling streams create swollen rivers as the excess water drains away. In the relatively flat central United States the rivers drain snowmelt slowly, with rises sometimes lingering for days or weeks as the water flows out of the system.

Thunderstorms can bring torrential rains as thunder cracks the skies open. The downpour first fills gutters and gullies, sometimes overwashing roads and bridges, and often rising rapidly. The dangerous flow can drag trees, cars, homes, and people into the watery web. It takes only six inches of rushing water to sweep a person off their feet and only twelve inches to move a car. Floods kill more people in the United States than any other weather hazard but heat.

Water is both life and death. It can lie placidly and nourish our bodies and souls. When it begins to move or rise, tranquility gives way to danger. The line when that happens is not always clear, however. Just about every driver has taken a car through a puddle, but the difference between a puddle and a flood is a gray area.

A stream can bubble safely under a tree, but when is its flow too swift for a person or a horse-drawn wagon to ford? An inch or two of rainfall can sponge into the ground or turn roads into rapids, depending on when and where it falls, how quickly, and how saturated the ground already was before it fell.

Flash floods, which cause most flooding deaths, are the rapid rises in water caused by downpours, dam bursts, ice jams, hurricanes, or other sources of water surges. The speed of the water itself, and the speed of the rise, factor as much as the depth of the water. A road that feels safe one moment can become questionable the next as puddles converge, then dangerous as water begins to flow over the road. While weather safety advice is to "turn around, don't drown" when approaching flooded roads, those already caught on the road may find themselves on an island in a rising sea, with no safe escape. Water can erode the roadbed, displacing chunks of concrete and asphalt or digging ditches across gravel; when those gaps are covered with just a few inches of water, they are no longer visible, and the road becomes a trap.

Then, too, the water that rose so quickly must drain through the river system. Rises can peak quickly, especially in the smaller streams and creeks, with water level measurements rising almost in a straight line as the runoff rushes into the stream, then move away downstream as quickly as when the water first arrived. The wider rivers rise later and more slowly, the shape of the water level rising as a rolling hill rather than a sharp peak. Summer sometimes brings a second season of river flooding, particularly during summers that are especially rainy over large tracts of the drainage areas.

Even placid floodwaters can drown farm fields, homes, roads, bridges, and businesses. The depth and persistence of the water, rather than its speed, compound the damage done. Roads soften, scour, and crumble. Homes wither and mold. Fields leach minerals and surrender topsoil, replaced by a blanket of silt and filth. When floodwaters threaten, the safest response is to get out of their way.

Fording a Rising River

Horse-drawn wagons often forded shallower streams and rivers as they moved across the country. Most rivers did not have bridges, and only the widest ones had costly ferries to transport wagons, their contents, and the horses that pulled them. The fords were always at least a little risky, as wagons could mire into the mud of the streambed, or float and tip over if the travelers did not estimate the river's depth carefully enough.

Rivers can flood from rains that fall tens or hundreds of miles away. The crest, or the tip of the peak of water level traveling down the stream, moves down one river, then into another as streams converge. The crest broadens and shallows as it travels downstream into wider and deeper rivers. A swell of water from upstream rains could take anywhere from hours to days to travel twenty miles downstream, depending on how deep and wide the river is and how much it drops in elevation, bringing floodwaters to areas that may have remained dry during the rainfall that caused the flood. Hydrologists can forecast how fast a crest will travel downstream, using computer models that account for the river's size and depth as well as the amount of water already in the river and the amount of rain that fell. The initial, swift rises at the beginning of a flood event are the ones that catch the most people off guard.

A steep rise is likely what the Ingalls family experienced when they forded a river somewhere between Kansas and Wisconsin. Wilder tells the story in *Little House on the Prairie* as if it occurred on the family's move south to Kansas, when Wilder herself was just a toddler. More likely it occurred on their return trip to Wisconsin, when she was about four years old and might remember the sensations of the wheels leaving the ground, floating free, bobbing, and finding solid ground again. Indeed, the *Pioneer Girl* manuscript places the dangerous river crossing on their return trip.[1] If Pa had attempted to ford a river with the horses and wagon, then the river likely was a shallow one, at least ordinarily. Everything of value to him was on that wagon, from his family to his fiddle to his horses.

The rise happened suddenly enough in Wilder's story that it more likely would have originated from a nearby rainfall, with a sharper peak, rather than a faraway rain with a broad and slow crest. Laura is lulled by the swaying wagon until Ma's frightened voice orders

1. Wilder, *Pioneer Girl: The Annotated Autobiography*, p. 21.

her and Mary to lie flat and still, and Pa leaps into the water to guide the horses. Laura peeks through the front of the wagon to see Pa and the horses swimming in deep water until the wheels and the horses' hooves strike creek bottom again. "I never saw a creek rise so fast in my life," Pa tells Ma on the other side.[2]

In *Little House on the Prairie*, the river crossing is close to their eventual homesite in Kansas, which would make a nearby river like the Verdigris River, just east of Independence, a potential culprit. In reality, the Ingalls family crossed multiple creeks and rivers between Kansas and Wisconsin, and Wilder's readers will never know which rising river nearly caused the disaster, just as Wilder herself almost certainly never knew. In *Pioneer Girl*, the river crossing episode occurs shortly after she notes that "we went on into Missouri."[3] It was a river with tracks from previous wagons fording the river, deep enough that the horses might have had to swim just a little, but not so wide that a rain-fed swell would have flattened out and caused a slow rise in the river. It had to be narrow enough to be responsive to those rains. Depending on their path, the exact details of which are not known, the Ingalls family could have crossed many rivers in Missouri: the Marais Des Cygnes, Osage, South Grand, and Blackwater Rivers, just to name a few, along with countless creeks.

The upshot is a cautionary tale about entering unknown waters and understanding upstream weather conditions. Wilder notes in *Pioneer Girl* before the crossing, "It was bad going for it rained making the roads muddy and the creeks and rivers high."[4] After the treacherous crossing, the family must have been scared into holding up on safe ground: "We waited then for the weather to settle and the waters to go down."[5]

Pa himself says in *Little House on the Prairie* that he would never have entered the waters if he had known how swiftly they were going to rise. Even in modern times, with flash flood warnings and rainfall forecasts, drivers can find themselves in similar danger, un-

2. Laura Ingalls Wilder, *Little House on the Prairie* (1935; Harper Trophy, 1971), p. 24.

3. Wilder, *Pioneer Girl: The Annotated Autobiography*, p. 18.

4. Ibid., p. 20.

5. Ibid., p. 21.

aware of how quickly the water could rise and sweep away their vehicle. While some choose to enter floodwaters in disregard for their danger, many others are there unintentionally, caught by surprise, just as Pa was. They may have been unaware of the danger, unfamiliar with where they were in relation to a place of danger, or caught in an event that was outside their previous experiences.

Cars and vehicles are not built to ford rivers, and the best place for them is parked on high ground, waiting for the weather to settle and the waters to recede. One who encounters a flooded road, especially with rising waters, should turn around and backtrack to safe ground, look for a safer way around, or better yet, wait until the threats have passed. Pet and Patty, the Ingalls family horses, may have swum through the flood, but iron horses do not fare so well.

Rainy Christmas

The Kansas winters were far milder than any others experienced by the Ingalls family during Wilder's childhood. Snowfall is a rarity in southern Kansas, perhaps just a few times per year. Temperatures can dip below freezing, but they typically do not linger there before rising to a thaw. The ground is almost never snow-covered for more than a day or two, and rivers run without ice throughout the year. All of these conditions are in stark contrast to the cold and snowy weather that the Ingalls family encountered in Wisconsin, Minnesota, South Dakota, and Iowa.

Wilder establishes a contrast between the Kansas winters and the northern ones in the weather she employs in her novels' settings, especially around Christmas. In *Little House on the Prairie*, a cold rain precedes Christmas, causing the Verdigris to swell. Ma weaves a heartwarming story for the girls about how Santa can reach children via wagon when there is not enough snow for his sleigh. But the river is too high for Pa to cross and visit nearby Independence, Kansas, presumably to purchase at least some Christmas candy for the family. As it turns out, their friendly neighbor, Mr. Edwards, plays Santa on foot for the family. As depicted in the book, he wades through a rainfall-swollen river to retrieve gifts for the family from Independence, braving the flooded waterway with a pack carried on his head. He delivers their gifts on Christmas morning with a wild story about how he met Santa in Independence, who asked him to deliver the presents to those Ingalls children because the river was just too flooded to cross.

A child reading the books may have overlooked or misunderstood why Pa and Ma Ingalls stand with tears in their eyes as Mr. Edwards tells his story to the children. Then again, even children can appreciate the generosity of a neighbor who risks his personal safety to make sure that his neighbor's children have Christmas presents. All the Ingallses have eyes filled with gratitude—the parents are grateful to Mr. Edwards, and the children to Santa, who gave them each a whole penny with their candy sticks and tin cups. The true identity of Mr. Edwards remains a mystery, and the act of generosity was attributed to a Mr. Brown in the *Pioneer Girl* manuscript.[6] Like other characters in the Little House books, Mr. Edwards is likely a composite character, as his appearances are threaded through three books in the series.[7]

The Ingalls family spent two Christmases, 1869 and 1870, on the Kansas prairie, but just one Christmas fit into the narrative arc of *Little House on the Prairie.* Weather records near Independence are scant for 1869 and 1870, as the nearest observing sites were in Lawrence, Kansas (some 150 miles to the north), and Harrisonville, Missouri (about 160 miles to the northeast, and for 1869 only).[8] Supplementing the observed records with reconstructed weather maps provides some confidence that the trends recorded at both Lawrence and Harrisonville were representative of the weather conditions near Independence, Kansas.[9]

December 1869 did begin on the mild side, with rain falling on four days from 4 to 11 December at both sites. It was heaviest on 10 December, with 0.27 inches of rain in Lawrence and 0.70 inches in Harrisonville. Conditions turned cooler around 14 December, with highs barely or not above freezing and lows dipping into the teens

6. Ibid., p. 16.

7. Robynne Elizabeth Miller and J. D. Rushmore, *Tennessee Wildcat: On the Trail of Laura Ingalls Wilder's Mr. Edwards* (Portland, Oregon: Practical Pioneer Press, 2023).

8. FORTS data.

9. 19th Century Reanalysis.

and twenties. Snow fell on 14–15 December, with a much heavier amount likely to have fallen at Harrisonville (on the order of several inches) than at Lawrence (around half an inch). Snow fell again on 21 December, with Harrisonville receiving the lighter half-inch while Lawrence received around five inches. The period from 17 to 24 December was particularly cold at both sites, with high temperatures remaining in the twenties and lows from near zero to the mid-teens in Harrisonville, while temperatures in Lawrence were just a few degrees warmer.

The weather in December 1870 was not all that different from December 1869, with a relatively mild first half of the month. In Lawrence, relatively wet weather did precede Christmas by about ten days, with 0.45 inches of liquid equivalent precipitation falling as about four to five inches of snow on 15 December. With a 2:00 p.m. temperature of 38°F but temperatures of 28°F at 7:00 a.m. and 31°F at 7:00 p.m., the temperature was borderline for rain or snow through the day—cold for the day, but not bone-chilling, and also not exactly mild, even for a family used to Wisconsin winters. The snow that fell may have melted that day or over the next couple of days. On 20 December, however, temperatures in Lawrence plunged to readings far below average. From 20 through 26 December, the warmest temperature was a rather cold 25°F on Christmas Day. Temperatures as measured at 7:00 a.m. during those seven days were as low as -10°F on Christmas Eve; again, Christmas was the "mildest" morning of that week, at 13°F.

The weather records, then, simply do not support Wilder's storytelling that the weather near Christmas was mild and rainy. In 1869, the liquid equivalent of precipitation between 25 November and 25 December was 1.29 inches, which is close to the long-term average for the area. About a third of that fell as snow, which is less likely to cause abrupt rises in rivers than a heavy rain. In 1870, the liquid equivalent precipitation between 22 November and 25 December was 0.86 inches, well below average. Most of that fell as snow, and except for the 0.45 inches on 15 December, it fell just a few hundredths to a tenth of an inch at a time. The rainfall pattern does not point to flooding conditions in either year, especially considering the very cold weather during the week before Christmas in both years.

So many of Wilder's weather stories were spot-on accurate that it feels jarring when a story does not reconcile with the records. The gap between Wilder's story and the recorded data is an uncomfortable space, one that reasonable explanations may fill, but nonetheless a puzzle that is missing too many pieces. One possible explanation is that Wilder herself did not recall the winters, leaving her to either rely on the memory of her family members in their long-past fireside talks of family lore or make up what happened and fictionalize it. If a friendly Mr. Edwards figure did indeed make a trip to town to bring presents for the family, perhaps the motivation was because it was too cold and snowy for Pa to travel, or perhaps it was simply a charitable gesture for a family with little money to spare. Wilder did explain away or neglect to mention other charitable gestures received by the family through the years; for every Christmas barrel that she did include, there was assistance from the state or a neighbor that she left out of the narrative. The gesture by "Mr. Edwards" is so heartfelt and warm that we want it to be true, even if the weather was not the reason. And if Pa and Ma had eyes filled with tears of gratitude they were a heartfelt response to his selfless act of bringing gifts to a family that, for whatever reason, could not get them on their own.

In a way, it is a relief that Mr. Edwards (or, perhaps more accurately, Mr. Brown) did not take a neck-deep plunge into a cold, flooded, fast-moving river in late December.[10]

Lake Pepin Ice Breakup

Ice is a solid, but it is not stable. It changes constantly. Even the thickest ice on a body of water in the middle of winter is not still. It can shift up or down, drift side to side, crack open, or freeze in an old crack. On most days, you can hear the ice sigh, groan, or whine. As ice slides, you can hear creaks, cracks, and pops. Nudged by the wind, the water currents, or sudden temperature changes, the ice moves audibly, like your arthritic relative standing up out of a chair with a groan and a popping knee.

Walking on the ice is not for the faint of heart. Walking safely

10. Miller and Rushmore, *Tennessee Wildcat*, p. 166.

means knowing how thick the ice is and the character of the ice itself. Is it clear and smooth, the depths of the lake visible beneath its icy window? Is it milky, with a bumpy surface as layers of snow have packed down and frozen? Is it slushy and soft in places? Is it cracked and shifting like tectonic plates as it breaks up, pushed by winds or currents? Despite its knee-trembling clarity, the clearest ice, sometimes known as "black" or "blue" ice, is the most solid and supportive. Milky "white" or "gray" ice can be stable, too, so long as it is intact, but every trapped air bubble that gives it the milky color is a potential weak spot. The snow on top can prevent the ice from hardening and adds weight to the load on the ice below.

Two inches of clear, or "black," ice on a still lake is the bare minimum that will support the weight of a person walking on foot cautiously, but four inches is safer. Five to seven inches will support a snowmobile, an ATV, or a loaded wagon. Eight to twelve inches will support a car.[11] As a rule, white ice should be at least twice as thick as black ice to support the same weight.[12] It is also important to remember that no ice is guaranteed to be safe.

In the days of covered-wagon and horseback travel, crossing rivers was no small matter. Without bridges across the widest rivers, travelers relied on ferries for deeper rivers and fords for shallower ones. To spare the expense of taking a wagon across a river on a ferry and the risk of fording a river, early settlers could sacrifice comfort and cross while ice still covered the rivers. The decision of when to cross became a balancing act of crossing early enough in the spring that the ice was still stable, but late enough that the cold weather would soon break.

The Ingalls family crossed Lake Pepin out of Wisconsin in 1869 and again in 1874, both times likely in the late winter or early spring.[13] Only a toddler in 1869, Wilder would have had no recol-

11. Tim Smalley and Kim Elverum, "Danger! Thin Ice," Minnesota Department of Natural Resources, 2010, https://files.dnr.state.mn.us/education_safety/ice/dangerthinice.pdf.

12. Jamie Orfald-Clarke, "6 Fundamentals of Ice Assessment," Boreal River Rescue, 18 Nov. 2020, https://rescue.borealriver.com/ice-safety-and-rescue/6-fundamentals-of-ice-assessment/.

13. Wilder, *Pioneer Girl: The Annotated Autobiography*, p. 46.

lection of the family's first crossing, and it is unlikely that she is retelling this crossing in *Pioneer Girl* or *Little House on the Prairie*. She does recall a crossing out of Wisconsin in *Little House on the Prairie*, creating narrative tension as the ice breaks the night after they cross the river. The memory in her *Pioneer Girl* manuscript is slightly safer, as the family crosses a little more in advance of the ice breakup, and it constructs a more thorough narrative for their late crossing.

As fortune would have it, the Minnesota Department of Natural Resources holds records back to 1843 of ice-out dates (the date of ice breakup) for several lakes in the state, including Lake Pepin—the oldest such record in the state.[14] The average ice-out date of the early part of the record (1843 through 1905) was 12 April. For both years that the Ingalls family crossed Lake Pepin, the ice-out date was later than average: 20 April 1869 and 22 April 1874.

The Ingalls family had planned their move in 1874, migrating into Minnesota with the family of Uncle Peter (Charles Ingalls's brother) and Aunt Eliza (Caroline Ingalls's sister), along with their children, Alice, Peter, and Ella. Several of the children, however, fell ill with scarlet fever. Laura was the last to recover, and the family departed before she had regained full health in order to make a safe crossing of Lake Pepin. Wilder recalls in *Pioneer Girl*:

> But when we were on the lake I heard the horses feet go "splash! splash!" in water and I was terrified. I thought we were breaking throug [*sic*] and would all be drowned in the lake under the ice, but Ma put her arm over me, drew me close and said it was all right. It was only a little water on top of the ice, so I went to sleep and waked up when they were putting me to bed in the hotel at Lake City, safely across the lake.[15]

Wilder also recalls that Pa bought her a book in Lake City the next morning to celebrate her birthday, on 7 February.[16] High tem-

14. Pepin ice-out data via Minnesota Department of Natural Resources, accessed 19 Feb. 2024, https://www.dnr.state.mn.us/lakefind/ice.html?id=25000100.

15. Wilder, *Pioneer Girl: The Annotated Autobiography*, p. 55.

16. Ibid., p. 59.

peratures did nudge above freezing sporadically on a few days in February 1874 in Madison, Wisconsin, and Minneapolis/St. Paul, Minnesota, two of the closest official weather stations to Pepin. Low temperatures remained below freezing, though, and the month lacked the kind of prolonged warming that would create water and slush on top of the ice cover. With low temperatures not reaching above freezing in February, it is more likely that her family did indeed cross late and prolong their departure as much as possible into spring to give the children time to recover. Temperatures in April did rise above freezing on several days, including minimum temperatures that stayed above freezing, which would support melting on top of the ice cover.

After a few days, the families traveled farther west, stopping over in a dugout home near a creek that soon flooded due to ice melt, again supporting a timeline of departure in April instead of February. Charles Ingalls filed his claim for the Walnut Grove land in New Ulm, Minnesota, on 26 June 1874, for land he had settled on 28 May, leaving ample time to travel from Pepin with an April departure.[17] Thus, the records allows us to conclude that the departure was indeed in spring, not in the heart of winter, pushed back as long as possible before the ice gave out on Lake Pepin. Wilder's recollection about a birthday present may have been asynchronous, or perhaps Charles Ingalls purchased the gift for a brave little girl who was still not feeling well when the family began a long journey west.

Spring Freshet

If it is frightening to be caught in a flood in a vehicle, whether powered by horse or engine, it is even more frightening to be bodily caught in the waters. Even in warm weather, waters sap heat from the body much faster than air does. People in floods quickly become cold and exhausted, clinging to whatever solid object holds them from being swept away.

Swelled by both snowmelt and rainfall, Plum Creek near the Ingalls family home rose out of its banks in Wilder's tales, described as Laura's naughty dip in the water in *On the Banks of Plum Creek*

17. Ibid., p. 6.

and quite differently when Laura seeks help for a very ill Ma in the *Pioneer Girl* manuscript.[18]

In *On the Banks of Plum Creek,* the spring freshet—a term used by some to describe stream and river flooding from snowmelt or heavy rain—brought water over the footbridge that crossed the creek and tantalizingly close to the dugout's front door. Laura, unable to resist the impulse to feel the water, dunks her toes and then her legs into the water. The current pulls her in and then pins her into its stream, and she clings to the footbridge as a lifeline, gathering the strength to pull herself out of the water.

If this episode really happened, Wilder was fortunate to have survived. Floodwaters are dangerous not only because of what the eyes see—the speed and depth of the current—but also because of what they do not see—the debris, sewage, animal waste, chemicals, and traps carried in or submerged beneath the water. A child is unlikely to be a strong enough swimmer to fight the current and stay afloat. Debris washed into the swift, swollen creek could have snagged her, loosening her grip on the footbridge or pinning her underwater. Once loose from the footbridge, she could have been swept into submerged logs or trees and pinned. Modern children also face the dangers of being swept into culverts and sewer drains.

In the *Pioneer Girl* manuscript, Wilder recalled the spring freshet shortly after Freddie was born, which would place it in the spring of 1876.[19] Ma was critically ill, and Pa sent Laura to ask the neighbors across the creek to fetch a doctor. The parents were so overwhelmed with Ma's illness that they forgot about the recent rains, and forgot that the creek had flooded and that no child could cross it. Driven by the urgency of Pa's request and Ma's critical illness, she waded into the water up to her knees to get to the footbridge. Fortunately, Laura spotted neighbor Mr. Nelson across the creek and was able to shout her message over the rushing water and retreat to safety. When she returned to the house, she quoted Pa with a "By Jinks!" when he saw how wet she was, likely only then realizing that he had

18. Ibid., p. 85.
19. Ibid., p. 83.

sent his daughter into a water trap.[20] A doctor arrived to care for Ma a day later, traveling to the house partly by boat.

The early months of 1876 brought precipitation amounts close to normal at Omaha, Des Moines, Minneapolis/St. Paul, and Duluth. However, the precipitation from January through May counted among the top ten wettest at many sites in the upper Mississippi River to Great Lakes region, such as La Crosse, Madison, Milwaukee, Dubuque, Moline, Lansing, and Detroit. Walnut Grove is geographically closer to the "near normal" group than the "well above normal" group, but it is also possible that the area of heavier rainfall from Wisconsin and Iowa could have included Walnut Grove while missing the other long-term sites around it.

Indeed, observers in neighboring Iowa reported several bouts of rain and thunderstorms that brought swaths of heavy rains across the state from February through May 1876.[21] Those under the swaths may have received an inch or two of rain during each round, while outside the swath, rainfall amounts quickly dropped to a quarter inch or less. Such swaths would explain pockets of wet conditions surrounded by sites with near-normal rainfall. The rains during the spring of 1876 tipped from below normal, to near normal, to wetter than normal from west to east across the Plains to the Midwest and Great Lakes. The spring was wettest in eastern Iowa and closer to normal in northwest Iowa, closer to Walnut Grove. It is rare to see a smooth gradient in rain for any one season from west to east, and it is reasonable for pockets of wetness and dryness both to be embedded in an area of near-normal precipitation. All of this leads to an informed conclusion, though only of moderate confidence, that it was probably wetter than usual near Walnut Grove in the spring of 1876, during the time of the spring freshet.

The decision to keep Ma's illness out of the narrative was an editorial one recommended by Rose Wilder Lane.[22] The book had

20. Ibid., p. 85.

21. John R. Shaffer, *Report of the Secretary of the Iowa State Agricultural Society, for the Year 1876* (Iowa State Agricultural Society, 1877), pp. 523–32, https://books.google.com/books?id=CPIWAAAAYAAJ&printsec=frontcover&source=gbs_ge_summary_r&cad=0#v=onepage&q&f=false.

22. LIW to RWL,13 June 1936, File 19, Rose Wilder Lane Papers, HHPL.

enough tragedy already, Lane told Wilder, and it was better not to mention Ma's sickness. Thus, the character Laura's walk through the freshet waters was motivated by childhood curiosity, rather than the urgent necessity that truly drove it. While the story emphasizing naughtiness may have sounded more pleasing for children, it does make considerably more sense that the only reason a responsible child such as Laura would have ventured into raging floodwaters was as an act of desperation to save her mother's life.

Rising Floods

While signals for changes in patterns for tornadoes and hurricanes are faint, like the ghostly images materializing in a developing Polaroid, the signals are clearer for changes in rainfall patterns in the Midwest and Great Lakes. From Minnesota and Iowa to Michigan and Ohio, precipitation has been increasing in winter and spring. These areas receive more precipitation now than they did in Wilder's day, part of the changing weather that those who take notice can observe in their own backyards and communities as climate changes.

Of course, even if the average trends toward wetter years, not every year is a wet one. Across Wisconsin, Minnesota, and Iowa, annual precipitation has increased by about 10 to 15 percent from the early twentieth century to the early twenty-first century.[23] For the winter months (December through February), though, the precipitation has increased by over 15 percent across much of the Midwest. Because winters are, in a word, cold in those states, some of that increased precipitation in winter falls as snow. At least for a while, when precipitation increases in the winter in colder climates, some of it is going to fall as snow, "juicing up" the snowfall totals. In a few decades, as wintertime temperatures continue to warm, those gains will melt away and the excess precipitation will fall more often as liquid instead of snow.

It is more than just the total rainfall that is on the rise in the Midwest and Great Lakes. The area is also seeing more of the total rain falling as heavy rain, the kind that is harder to drain away. The

23. "Ch. 2: Climate Trends," Fig. 2.4, in *Fifth National Climate Assessment*, https://nca2023.globalchange.gov/chapter/2/#fig-2-4.

flooding kind. Again, it doesn't mean that a heavy rain happens with every rainfall or that floods happen every year. It just means that they are happening more often now than they used to in Wilder's days. Total summertime precipitation in the Midwest and Great Lakes has not changed much, at least in average amounts calculated for the season. If roughly the same amount of rain falls, but it falls more often as heavy rain, then longer gaps of dry weather must be occurring between those heavy rains.

Those swings in precipitation are hard on farmers trying to plan for watering their crops. In Iowa and surrounding states, farmers rely on new technologies, tested by scientists, to better manage water excess and shortage. One technique they use is drain tiling. While the world "tiling" evokes an image of laying squares of material to cover the ground, "tile" in the world of farming is a series of drainage pipes laid underground beneath a field to aid drainage. The tiles lower the water table in fields, which allows crops to develop deeper root systems that foster healthier growth and can offset minor flooding. They are not without drawbacks, though, as they also shepherd nitrates and other pollutants into waterways.[24] Researchers are working on new techniques and technologies to minimize the nitrate seepage and improve the water quality around the drained fields.[25] In many communities, farmers are leading the efforts to work with the shifts in weather they have observed that affect their crops.

The Missouri River has a history of flooding after wet winters

24. Mark B. David, Laurie E. Drinkwater, and Gregory F. McIsaac, "Sources of Nitrate Yields in the Mississippi River Basin," *Journal of Environmental Quality* 39, no. 5 (September 2010): 1657–67, https://doi.org/10.2134/jeq2010.0115 (also noted in https://phys.org/news/2010-09-tile-drainage-nitrate-loss.html, accessed 25 Feb. 2024); and Eileen Kladivko and Laura C. Bowling, "Long-term Impacts of Drain Spacing, Crop Management, and Weather on Nitrate Leaching to Subsurface Drains," *Journal of Environmental Quality* 50, no. 3 (May/June 2021): 627–38, https://doi.org/10.1002/jeq2.20215.

25. Chris Hay, "Chapter 31: Reducing Nitrate Losses from Drained Lands," in D. E. Clay, C. G. Carlson, S. A. Clay, and E. Byamukama, eds., *iGrow Corn: Best Management Practices* (South Dakota State University, 2016), https://extension.sdstate.edu/sites/default/files/2019-09/S-0003-31-Corn.pdf, accessed 25 Feb. 2024.

either near the river or upstream in its mountain headwaters. Similar to many rivers, a flood plain (the flat area near the river itself) surrounds the current stream channel. This flood plain has been washed over with flooding for centuries before humans brought in dams, levees, and other control measures. The floods were both a blessing and a curse, as they restored nutrients to the soil but also could drown things growing in their path.

The engineering that has brought the dam and reservoir systems and levee protections now controls the river's flow, but not perfectly. No human engineering can stop all floods. The system can manage high flows and hold back some water to let it drain more slowly, and it can keep some water in storage to maintain water in the stream in dry years. But no system is perfect. Floods will occur no matter how much structure is built, even if the humans in the process make the best possible decisions with the information they have at the time. Rivers are designed in nature to flood into their flood plains, even if humans and other animals wish to occupy the space.

Cities built along the rivers in the Midwest and Great Lakes bear the brunt of prolonged river flooding. For many years, floods in the spring of 1881 held the records, but floods in the 1940s and 1950s, including 1943, 1944, and 1952, again destroyed parts of cities like Omaha, Council Bluffs, Sioux City, and Kansas City.[26] The Great Flood of 1993 felt like an unmatched flooding disaster at the time, with extensive flooding across nine states along the banks of the Mississippi and Missouri Rivers and their tributaries.[27] Dozens of

26. "Historic Floods on the Missouri River: Fighting the Big Muddy in Nebraska," Nebraska Department of Natural Resources (October 2008), accessed 25 Feb. 2024, https://semspub.epa.gov/work/07/30022840.pdf and https://web.archive.org/web/20080603100822/http://www.dnr.ne.gov/floodplain/mitigation/mofloods.html; *see also* Joseph Vederman, "Meteorology of the Mid-West Floods of May, 1943," *Bulletin of the American Meteorological Society* 26 (October 1945); Todd Kerstetter, "'The Worst Floods in History': Federal Government and the Floods of 1944 in the Elkhorn River Basin," *Great Plains Quarterly* 21 (Summer 2001): 179–92, https://digitalcommons.unl.edu/greatplainsquarterly/2260/, accessed 25 Feb. 2024; and Ginalie Swaim, "Against All Odds: The 1952 Missouri River Flood at Council Bluffs and Omaha," *Iowa Heritage Illustrated* 91, no. 4 (2010): 124–33, https://doi.org/10.17077/1088-5943.1499.

27. Kenneth E. Kunkel, Stanley A. Changnon, and James R. Angel, "Climatic Aspects of the 1993 Upper Mississippi River Basin Flood," *Bulletin of the Ameri-*

small towns along their banks suffered for months underwater, as did parts of large Midwestern cities like St. Louis, Kansas City, Omaha, Des Moines, and the Quad Cities (Davenport, Bettendorf, Rock Island, and Moline) in Iowa and Illinois. Major highways and bridges closed for months, as did major rail lines, hampering travel and the transportation of goods across the country. Then the floods came again in 2008, and again in 2011, and again in 2019, and again in 2024.[28]

Cities do learn from the floods and adapt steps to become more resilient to future floods. Cedar Rapids, Iowa, was devastated by the flood of 2008 along the Iowa River.[29] Following the disaster, Cedar

---

can Meteorological Society 75, no. 5 (1994): 811–22, https://doi.org/10.1175/1520-0477(1994)075<0811:CAOTUM>2.0.CO;2; and Gerald E. Galloway, Jr., "Learning from the Mississippi Flood of 1993: Impacts, Management Issues, and Areas for Research," U.S.-Italy Research Workshop on the Hydrometeorology, Impacts, and Management of Extreme Floods, Perugia, Italy, Nov. 1995, accessed 25 Feb. 2024, www.diam.unige.it/idromet/workshop/index.html.

28. Stanley A. Changnon, *Railroads and Weather: From Fogs to Floods and Heat to Hurricanes* (American Meteorological Society, 2006); James A. Smith, Mary Lynn Baeck, Gabriele Villarini, Daniel B. Wright, and Witold Krajewski, "Extreme Flood Response: The June 2008 Flooding in Iowa," *Journal of Hydrometeorology* 14, no. 6 (2013): 1810–25, https://doi.org/10.1175/JHM-D-12-0191.1; Robert R. Holmes, Jr., Todd A. Koenig, and Krista A. Karstensen, *Flooding in the United States Midwest, 2008* (U.S. Geological Survey, 2010), accessed 25 Feb. 2024, https://pubs.usgs.gov/pp/1775/pdf/pp1775.pdf; National Weather Service, *Service Assessment: The Missouri/Souris River Floods of May-August 2011* (U.S. Department of Commerce/National Oceanic at Atmospheric Administration, May 2012), accessed 25 Feb. 2024, https://www.weather.gov/media/publications/assessments/Missouri_floods11.pdf; Lily L. Kraft, Gabriele Villarini, and Jeffrey Czajkowski, "Characterizing the 2019 Midwest Flood: A Hydrologic and Socioeconomic Perspective," *Weather, Climate, and Society* 15, no. 3 (2023): 603–17, https://doi.org/10.1175/WCAS-D-22-0065.1; Andrew Nystrom, *Leading the Charge Against the June 2024 Missouri River Floods* (U.S. Army Corps of Engineers, 2024).

29. U.S. Army Corps of Engineers Institute for Water Resources, "Cedar Rapids Flood Risk Management: A Case Study in Disaster Recovery," 2018-R-2, U.S. Army Corps of Engineers, 2018, accessed 25 Feb. 2024, https://asfpm-library.s3-us-west-2.amazonaws.com/Website/CON/B4-Jacobs; and Juan Fung, "Resilient Cedar Rapids: Rebuilding Better Pays Off," *Taking Measure: Just a Standard Blog*, 12 Aug. 2020, https://www.nist.gov/blogs/taking-measure/resilient-cedar-rapids-rebuilding-better-pays.

Rapids citizens contributed to the city's resilience plan, focusing on flood control infrastructure and neighborhood revitalization to attract new residents. Instead of focusing only on disaster recovery, the city and its citizens chose to create a resilient city, providing benefits even if a disastrous flood like 2008 never happens again. The community learned from its past, from what feels like the worst that could happen, to protect their safety and improve their city for the future.

One option for those who live near flood plains is to pay for flood insurance. While homeowner's insurance is required for everyone with a mortgage, it does not include coverage for flood damage, a fact that often comes as a gut-wrenching shock to homeowners in the wake of flood disasters. Flood insurance is mandatory for anyone who lives in a "100-year" flood plain—that is, an area with at least a 1 percent chance of flooding every year. It is available to everyone who wants it, though. And it is expensive. But so is losing everything in a flood.

Flooded rivers, and flash flooding over land, can originate from many sources. Spring snow melt can fill rivers and low-lying ditches, with high river flows lingering for days, weeks, or even months in the worst cases. River ice jams can send water quickly up and over the banks. Thunderstorms overhead can bring water up rapidly over streets and bridges, creating danger in normally dry places. Storms farther away can send a wash of water downstream, causing a river surge far from the source. While just one storm can produce flooding, repeated washes of many storms over many days or weeks can inundate large swaths of land, as they did in the summer of 1993 in the middle of the United States. Floods also stem from hurricanes and tropical storms, unusually high coastal tides, dam breaches, and even the sudden heat of a volcanic eruption melting mountain glacial ice.

The threat from floodwaters can mask itself in slow rises to river levels, or it can arrive in a deluge. The power of water can stem from its force, its persistence, its depth. A safe river crossing can quickly become dangerous, whether on foot, on a wagon, or in a car.

A crossing can stay unsafe for minutes, hours, or months. The water decides how long it will stay.

At least the water does show itself, even if it masks its danger in mysterious depths. It takes careful attention to see the danger, especially when it travels in from out of town or seeps in slowly. The signs are there, though, from reports of heavy rains or snow melt to floodwaters lapping at the edges of roads and bridges. Reacting to flood threats requires some trust in reports from far away and in predictions from hydrologists, as well as knowing the path to safety.

8

All Turns to Dust and Drought

As a young meteorologist early in my career, I joined my colleague and mentor Ray Wolf in our small office conference room to deliver a press conference. It was July 2005, and hot and dry weather was baking the corn and soybean fields of Iowa and Illinois. Representatives from the U.S. Department of Agriculture and Iowa State University Extension Service joined us so that we could combine weather, climate, and agriculture information, getting a unified message about the weather and its impacts out for the farmers in our area. The university extension crop specialist stood up and grimly gave the state of the corn and soybean crops and their stunted growth. He said that the last thing they needed was hot, windy weather.

My job was to give the weather portion of the briefing, with the forecast for the upcoming week—hot, dry, and windy. "Don't shoot the messenger!" I quipped as I walked up to the podium to deliver the bad news. I learned a valuable lesson that day: everything said at a press conference is fair game, even when it isn't at the podium. "Don't Shoot The Messenger," the newspaper headline read the next day.[1] Lesson learned.

As part of my immersion in drought, I also walked

1. Thomas Geyer, "Q-C Braces for Heat Wave," *Quad City Times*,

a dusty field that summer. As we stepped to the edge of the field, I was already hoping that the corn in the middle was not nearly as stressed as what I saw near the road. Edge rows are sometimes not as healthy as interior rows, but these were particularly stunted and browned. It was mid-July and the stalks were nowhere close to putting up tassels, when they should have already been sprouting bushy tops. As I pushed farther into the field, I noted that the edges of the sharp corn leaves were yellowing when they should be a brilliant Kelly green. Leaves near the top had a slight inward curl. The interior rows were tasseling, primed for pollination, but the tassels were crisp and thin. At a time when the corn needs the most water for healthy growth, it was getting next to none.

The fields were not irrigated, being in a climate that is wet enough not to need it most years. Irrigation is expensive, sometimes messy, and not needed in many years, so dryland farmers take their chances. Years like 2005 make the label "dryland farming" feel especially wry.

The flash drought in 2005 cost around $2.5 billion, mostly in staggering corn and soybean crop losses.[2] Fast-developing drought materialized again in 2012, much larger in area, on the heels of an especially warm and somewhat dry spring. That drought, too, caused expensive and expansive losses to corn and soybean harvests, as well as to sorghum and other crops. Losses due to drought in 2012 across the U.S. topped $40 billion.[3]

Cracked, bare ground. Gritty dust in the air, filtering into homes, sticking to the inside of a nose, crunching between teeth. Shimmering, waggling waves of heat rising from the ground, distorting the view of the distance. Looming, ground-scouring black clouds of dirt consuming swaths of ranch land and Model Ts. The iconic images of drought, like these from the Dust Bowl era of the 1930s in the

---

22 July 2005, https://qctimes.com/news/local/q-c-braces-for-heat-wave/article_4c7e18cf-d434-5108-8c9c-c3ceobf19c11.html.

2. "U.S. Billion-Dollar Weather and Climate Disasters," NOAA National Centers for Environmental Information (NCEI), 2024, https://www.ncei.noaa.gov/access/billions/, DOI:10.25921/stkw-7w73.

3. Ibid.

central United States, often define our perception of what drought is and what it does to the ground and the air around us.

The look of drought can come in many forms, just as drought itself can come in many forms. A flash drought, for example, strikes quickly in mid-summer as sudden hot, dry air bakes the ground like an oven.[4] Flash drought can turn a healthy corn crop into a faltering one in a matter of just weeks. The growing corn plants, which emerged healthy and green, hit the brakes. Their wide, flat green leaves curl in retreat, their edges browned and crisped with the singe of suddenly hot, dry air. The soil below dries, hardens, and cracks. Farm ponds shrink to puddles. A flash drought can depart as quickly as it arrives, with rains restoring the topsoil moisture and refilling farm ponds. But the damage is done, as flash droughts in mid-summer often strike at critical reproductive times for corn and soybean plants. Or it can be the start of a drought that lasts for months or years, the first snowball of an avalanche.

Often, droughts are more insipid, their beginnings and endings as ill-defined as the edge of the wind. Droughts commonly stretch from months to a year or longer. Droughts in intensive agriculture swaths like the Corn Belt, the Wheat Belt, and the Cotton Belt hamper the year's harvest, activating relief from the Department of Agriculture and its many assistance programs. The drying vegetation can turn tree and plant debris into kindling, increasing the risk from wildfires. Homeowners may be inconvenienced by restrictions on watering lawns and washing cars, while ranchers may need to sell livestock they can no longer afford to feed and water. The depleted surface water can strain farming and ranching operations, especially those that do not have deep irrigation wells. With years of

4. Caily Schwartz, Tonya Haigh, Mark Svoboda, and Madeline Goebel, "How is Flash Drought Understood?—Experts' Definitions and Decision-Makers' Perceptions," *Weather, Climate, and Society* 15, no. 4 (2023): 909–24, https://doi.org/10.1175/WCAS-D-22-0142.1; Mark Svoboda, et al.,"The Drought Monitor," *Bulletin of the American Meteorological Society* 83 (2002): 1181–90, https://doi.org/10.1175/1520-0477-83.8.1181; and Jason A. Otkin, Mark Svoboda, Eric D. Hunt, Trent W. Ford, Martha C. Anderson, Christopher Hain, and Jeffrey B. Basara, "Flash Droughts: A Review and Assessment of the Challenges Imposed by Rapid-Onset Droughts in the United States," *Bulletin of the American Meteorological Society* 99 (2018): 911–19, https://doi.org/10.1175/BAMS-D-17-0149.1.

dry weather, reservoirs can empty from deep lakes into puddles and streams, with docks and moorings left high and dry and boat ramps ending in rocks and gravel. Bathtub rings of mineral deposits from depleted reservoirs like Lake Powell and Lake Mead in Arizona cling to the rocks at their edges, a glaring reminder of the depth of water in flush years. The effects aren't all recreational, though, as fish and bird species suffer from lack of water, too, with ripple effects both up and down the food chain.

Droughts can last for years, even decades, especially in vulnerable places like California, the southwest, and (rarely but increasingly) parts of the Plains. A wet month here or there is not nearly enough to offset the years of dryness, merely causing a brief flush of green before the hot and dry weather returns to burn it up again. In the longest droughts, water far below the surface is depleted, too. The groundwater level drops, sometimes falling below the depth to which wells have been dug. Swaths of forest can wither and die altogether, dry for too long to recover.

As stark as some images of drought can be, it doesn't always look so dry. Surface conditions can recover, either temporarily or on a path to long-term wetness, while drought lurks on below the ground. Groundwater and reservoirs can remain low after a wet period, especially if the drought has been festering for years. Sudden, heavy rainfall in drought zones can still create flash flooding, the rain falling too quickly for the soil to properly absorb it, running off instead of restoring deep moisture. Droughts in normally wet climates like the northeastern United States look more like an early fall's brown leaves and grass than a Dust Bowl panorama of black blizzards and drifting dirt.

Defining Drought

Compared to some other weather definitions, drought is defined rather loosely, as a period of abnormally dry weather over short or long durations that causes a hydrological imbalance that has an impact.[5] In other words, drought means not having enough water, for

5. Donald A. Wilhite and Michael H. Glantz, "Understanding the Drought Phenomenon: The Role of Definitions," *Water International* 10 (1985): 111–20,

some period of time, compared to the usual balance of a location. A drought can last for a month or for decades, over small swaths or large. Heat is not a requirement for creating drought, but it does reinforce the drying process by evaporating moisture from the ground. Because the drier ground can heat more quickly, drought and heat can become symbiotic.

Droughts are sometimes labeled by a combination of their causes, impacts, and duration. A flash drought, as discussed above, arrives rapidly—usually on a summer heat wave—and can retreat later in the same season. A meteorological drought can be traced to a lack of precipitation for a month, a season, maybe a year. A hydrological drought often lags the onset of dry weather conditions, when the long-term shortages affect groundwater, reservoirs, and river levels. Agricultural drought is a dry spell that hampers crops in a location at that time and can include everything from the short-lived flash droughts to longer-lasting meteorological droughts during the crucial growing season.

Socioeconomic drought, wherein human demand for water exceeds the available supply, adds another layer of complexity: human behavior. While lack of rain is one factor, it is not usually the only one. Water management practices that do not keep enough water in reservoirs, over-allocated rivers in systems that define river flow by treaty and compromises, and over-irrigation that has dropped groundwater below reachable levels are all ways that humans compound water shortages in dry years.

As the climate warms, the added heat in the system will cause more frequent droughts in areas that already lean warm and dry, such as the southwestern United States, and in areas on the boundary of dry climates, like the Rocky Mountains and High Plains. Drought can disrupt the water storage systems needed in the western United States to sustain the cities that have been built there as less snow falls and less snowmelt runs off through the rivers.

In the modern era, meteorologists and climatologists track

---

https://doi.org/10.1080/02508068508686328; "Drought," Glossary of Meteorology, American Meteorological Society, last modified 27 Mar. 2024, https://glossary.ametsoc.org/wiki/drought.

drought by gathering records of precipitation, temperatures, and vegetative health via satellite-observed greenness. They also gather stories of impacts due to dryness—the words of those who live there and have some perspective, such as a rancher, a farmer, an extension agent who reaches many farmers, a water resource manager, or a host of others who pay attention to the lack of water.

Droughts were even more loosely defined in the mid- to late nineteenth century. Homesteaders had comparatively brief lived experiences or observations in their new home areas, and they were unlikely to deeply consider the input of the Lakotas and other tribal Plains residents who did. Any relative dryness could have been labeled a "drought" in their perception, even if it would not be one by today's standards or with a longer perspective of weather and climate observations. American Indian counting calendars give evidence that drought, or lack of water, was important enough to record.[6]

Droughts are a normal part of the "average weather" in many locations, just as tornadoes, hurricanes, and floods can all be a part of the expected range of weather. It is normal to have periodic droughts in the Plains, occurring once or twice per decade and lasting for maybe a year or two, while droughts are less frequent in the Midwest and Great Lakes.

However defined and declared, drought is among the most expensive weather-related disasters in the United States, costing millions to billions of dollars per year depending on extent and severity. Farmers and ranchers bear the brunt of the cost, though many have crop insurance through the Department of Agriculture to help absorb the impact, shifting the brunt to the federal government. Costs can accrue in the efforts to seek additional water supplies, move infrastructure (like boat ramps) to accommodate lower water levels, and battle drought-fueled wildfires, as just a small set of expense examples. Drought is also connected to an increase in mental health problems in hard-hit communities.[7]

6. Greene and Thornton, *Year the Stars Fell*, p. 128.

7. Jesse D. Berman, Marizen R. Ramirez, Jesse E. Bell, Rocky Bilotta, Fredric Gerr, and Nathan B. Fethke, "The Association between Drought Conditions and Increased Occupational Psychosocial Stress among U.S. Farmers: An Occupational Cohort Study," *Science of the Total Environment* 798 (2021).

Dry Days on Plum Creek

From the banks of Plum Creek arises the first thread woven through the books of the recurring theme of drought and its impact on the Ingalls and Wilder families. In *On the Banks of Plum Creek*, drought and grasshopper weather are linked, both driving crop losses just when the Ingalls family needs to bring in a good crop and get ahead with their finances. Wilder wrote of Plum Creek running nearly dry, of a string of hot days, of a neighbor who had "never known such a dry, sunny summer."[8]

The records from the official weather observing sites provide information about rain during the Ingalls family's years near Walnut Grove, Minnesota (1874–1876 and 1877–1879), but rain is notoriously fickle, often falling heavily in swaths just miles from patches that remain dry. This is especially true of summertime rainfall, which tends to fall under thunderstorms with smaller areas of coverage than the more broad-brush cold-season weather systems. A map with a bird's-eye view of rainfall can be valuable to understanding the pattern of rainfall coverage, but one is not available from the 1870s, before radar and satellite data could help fill the gaps between observing sites and before there was enough data in more sparsely populated areas to fill in a map with more than just a few points across wide expanses of land.

Confidence in scientific conclusions is always a little lower when the records do not include direct measurements. Even so, scientists can make well-reasoned conclusions that stand up until additional records, stories, and research can fill the gaps and perhaps suggest different conclusions. This is part of the scientific process: gathering the best data available to make scientific hypotheses, then adjusting the hypothesis when additional, better data suggests it is needed.

In any case, dry conditions prevailed across most of the central Plains in the summers of 1874 to 1876, with the driest areas extending across Kansas, Nebraska, and South Dakota. The core of drought conditions reveals itself clearly in maps of atmospheric moisture from computer models that reconstruct the atmosphere. Its edge is hazy, probably shifting from year to year as rains en-

8. Wilder, *On the Banks of Plum Creek*, pp. 206, 213–19, 189.

croached on its edges, then reclaiming the fringes in drier months. If southern Minnesota experienced drought conditions, it was on the edges of the main swath centered in Kansas and Nebraska.

At Minneapolis/St. Paul, 117 miles east of Walnut Grove, the years of the Ingalls family's first stay in Walnut Grove were the fifteenth wettest (1874), forty-fifth wettest (1875), and thirty-eighth driest (1876)—not exactly an extended drought. Even comparing to the climatology of 1871–1900, closer to the Ingalls family's residence in the region, the years were wet, wet again, and dry, respectively. Omaha, southwest of Walnut Grove, is upstream of the latter in prevailing atmospheric flow and could be more representative of the conditions experienced there during the drought of the 1870s, even if Minneapolis/St. Paul is closer. In Omaha, those same three years were the fifty-first driest (1874), seventh wettest (1875), and forty-fifth wettest (1876), still not reflecting widespread drought, but rather more indicative of up-and-down rainfall totals over those three years in the Midwest.

But drought is defined by its impacts as much as it is by precipitation totals. If creeks that don't normally run dry do so in the summer, and if crops that usually grow well wither in a field, then dry weather did have some kind of impact. Just like the textbook definition, dry weather with an impact is a drought.

Persistent drought over a wide segment of North America is often connected to global weather patterns, and the 1870s drought is no exception. Looking back to "The Boy Child" (Chapter 4), La Niña, the colder sister of El Niño, emerges as a piece of the puzzle of weather patterns that shaped the Ingalls family's experience. During La Niña, the trade winds in the Pacific Ocean near the equator increase and push the warm surface water into the western Pacific Ocean, leaving colder waters off the coast of South America and pushing the genesis region for tropical thunderstorms farther away to the west. The downstream jet stream pattern becomes wavier, often linking the more southern subtropical jet with the more northern polar jet. While the jet waves across the northern United States, dragging colder air with it, the southern Plains (especially Texas, Oklahoma, and eastern New Mexico) can be left high and dry.

La Niña can persist through summer months and across multiple consecutive winters. When La Niña persists or dominates for several years in a row, the drought areas that originate in the southern Plains reaches of Texas, Oklahoma, and New Mexico can spread to adjacent areas of Colorado, Kansas, Nebraska, Missouri, and beyond. This very phenomenon occurred in 2011–2012 and 2020–2022 as drought developed in the southern Plains and spread northward into an intense central United States drought, along with summer heat waves in 2012.

La Niña dominated weather systems from 1872 to 1875, with conditions borderline to La Niña definition thresholds in 1875–1876.[9] With the worst drought conditions centered in Kansas, pockets of drought developed in Texas, Oklahoma, Kansas, Colorado, and even in Wyoming and Nebraska through the decade. Temperatures were above average, too, relative to 1871–1900 averages. The upper Midwest and Great Lakes regions, where the Ingalls family resided, saw more swings between periods of wetter weather and periods of drier weather.

The intermittent drought conditions around Walnut Grove in the 1870s were enough to push a delicate balance of survival to the ground for the Ingalls family. Strained for money and struggling to raise crops under the siege of both drought and grasshoppers, they left farming after three difficult years.

"Rain Follows the Plow"

Though scientists have thought for centuries that humans can change the climate around them, they have not always had a clear picture of how. Few examples of misunderstanding the interplay between people and the climate around them are more egregious than the homestead-era settlement of the Great Plains.

The federal government and railroad companies alike sold the central and High Plains "frontier" to would-be settlers with palpable exaggeration and optimism. Drawing conclusions based on

9. Brian McNoldy, "Historic Oceanic Niño Index and Niño 3.4 Data," Rosenstiel School of Marine and Atmospheric Science, University of Miami, accessed 22 Mar. 2024, https://bmcnoldy.earth.miami.edu/tropics/oni/.

incomplete information, naturalist Samuel H. Aughey published a scientific paper in 1880 that gave credit for a recent wet period to the overturning of prairie sod.[10] A year later, scientist and journalist Charles Dana Wilber published four highly influential words in his book *The Great Valleys and Prairies of Nebraska and the Northwest*: "Rain follows the plow."[11] Both Aughey and Wilber extolled Nebraska's climate with far more virtue than it contained, caught up in boosterism and incomplete information about Nebraska's climate history and future. The catchy phrase, however, clung to the lips of settlers and speculators alike. Boisterous brochures from railroad companies promised "limpid waters" in "Nebraska, garden of the West" and "pure and abundant water" in southwest Kansas.[12]

The concept behind "rain follows the plow" was that the act of farming the land (plowing up the sod) would create a presumably symbiotic cycle. Aughey believed that plowing the land and exposing the soil would cause it to absorb more water, which it could then release into the atmosphere where the water would more readily fall as rain. Scientists at the time were not unanimous in their agreement, as others pointed out that the rainfall could be cyclical between wet and dry periods, or that the extra trees that had been planted by the settlers already present in the lands had caused the

10. Samuel Aughey, *Sketches of the Physical Geography and Geology of Nebraska* (Omaha: Daily Republican Book and Job Office, 1881), https://books.google.com/books?id=J20tAAAAYAAJ&pg=PP7&source=gbs_selected_pages&cad=1#v=onepage&q&f=false.

11. Charles Dana Wilber, *The Great Valleys and Prairies of Nebraska and the Northwest* (Omaha: Daily Republican Print, 1881), p. 68, https://www.google.com/books/edition/The_Great_Valleys_and_Prairies_of_Nebras/tAQUaD7pnhsC?hl=en&gbpv=1&printsec=frontcover.

12. "Thinking West: Advertisements for Western Land," *El Portal Journal*, 26 Nov. 2014, https://elportaljournal.com/2014/11/26/thinking-west-advertisements-for-western-land/; Kansas Historical Society, "Read Kansas! Middle School—M-41 The Great American Desert," accessed 22 Mar. 2024, https://www.kshs.org/p/read-kansas-middle-school-m-41-the-great-american-desert/17347; Granger Historical Picture Archive, "Railway Poster, 1870s. A Chicago and North-Western Railway poster of the 1870s promoting free homestead lands in the Dakota Territory," accessed 22 Mar. 2024, https://www.granger.com/results.asp?inline=true&image=0029123&wwwflag=4&itemx=5.

wet period. Aughey couldn't have been farther from the truth, as settlers would later learn in the Dust Bowl years.[13]

One kernel of truth at the center of the claim is that human modifications to the landscape can indeed alter the climatology near it. Converting native grassland to agricultural land can increase the dewpoint temperature—that is, the humidity—over and near that land. Evapotranspiration is the exhale of water into the atmosphere by plants. Plants like corn and soybeans draw moisture from the ground through their roots into their leaves, stems, and fruits. The water evaporates into the surrounding air through the leaves, joining forces with neighboring water molecules to humidify the air. If there were just a few plants here and there, it would take highly sensitive instruments to measure the difference in moisture. With millions of acres of densely planted agriculture in the Corn Belt, the collective exhale of all those crops puts a measurable amount of moisture into the atmosphere.

The added humidity is especially notable in areas of irrigated agriculture, where giant pumps pull millions of gallons of water out of aquifers below the ground and spray it across the corn and soybean fields that color the plains green through the summer. A tiny fraction of this water filters through the soil back into the aquifer, but the thirsty roots of the corn and soybean plants drink most of it. Enough water enters the air to create a different climate in the irrigated land than in the non-irrigated surroundings. The effects are strongest in the heart of the planted corn and soybean fields, but one need not stand in a field to feel the heat. The moisture follows the winds, mixing around to blanket the region and causing higher heat indices, possibly taking an oppressive day to a dangerous one. Higher dew point temperatures also can prevent tempera-

13. Gary D. Libecap and Zeynep Kocabiyik Hansen, "'Rain Follows the Plow' and Dryfarming Doctrine: The Climate Information Problem and Homestead Failure in the Upper Great Plains, 1890–1925," *Journal of Economic History* 62, no. 1 (2002): 86–120, http://www.jstor.org/stable/2697973; Cary J. Mock, "Rainfall in the Garden of the United States Great Plains, 1870–1889," *Climatic Change* 44 (2000): 173–95, https://doi.org/10.1023/A:1005570827123; and Henry Nash Smith, "Rain Follows the Plow: The Notion of Increased Rainfall for the Great Plains, 1844–1880," *Huntington Library Quarterly* 10, no. 2 (1947): 169–93, https://doi.org/10.2307/3815643.

tures from cooling at night, leaving little respite from the heat. The moisture can add juice to summertime showers and thunderstorms in the Midwest.[14]

The interaction of humans, land cover, and climate becomes even more entangled when considering tree claims under the Timber Culture Act of 1873. Again based on a kernel of truth that areas with forests are cooler and less windy than the same areas after they are deforested, the federal government distributed tree claims, encouraging settlers to plant tracts of trees across the Plains to help cool the temperatures and break the winds.

The practice of planting shelterbelts as windbreaks and habitats to support soil conservation in the Plains began during the Dust Bowl era and is still encouraged to this day. Many 1930s shelterbelts are either withering due to age or neglect or are being removed to make more room for crops, as the hard lessons of the Dust Bowl fade from memory.

Perhaps the biggest challenge to tree claims in the Plains, however, is that swaths of forested land are not native to the region and thus are not naturally sustainable. With great care, irrigation, and some luck, a tract of planted trees may survive. As most tree claim settlers found, though, planting forests on the prairie was doomed to failure from the outset. The practice was abandoned in 1891 due to both the difficulties of growing trees in an arid climate and widespread fraud associated with speculation using the Timber Culture Act.[15] Almanzo Wilder had a tree claim himself, planting cottonwood, willow, and boxelder trees.[16] While all of those species are native to the Dakota prairies, they tend to grow near bodies of water, rather than in neat rows and groves. Few tracts survived for long.

14. David J. Kellenbenz, Thomas J. Grafenauer, and Jonathan M. Davies, "The North Dakota Tornadic Supercells of 18 July 2004: Issues Concerning High LCL Heights and Evapotranspiration," *Weather and Forecasting* 22, no. 6 (2007): 1200–1213, https://doi.org/10.1175/2007WAF2006109.1.

15. C. Barron McIntosh, "Use and Abuse of the Timber Culture Act," *Annals of the Association of American Geographers* 65, no. 3 (1975): 347–62, http://www.jstor.org/stable/2561886.

16. Marta McDowell, *The World of Laura Ingalls Wilder: The Frontier Landscapes that Inspired the Little House Books* (Portland, Oregon: Timber Press, 2017), p. 204.

One study found that of forty tree claims randomly selected in Nebraska, just one remained treed by 1974.[17]

In *These Happy Golden Years,* Almanzo succinctly explains to Laura the concept of tree claims: the government officials had mapped where they wanted trees planted.[18] Both Almanzo and Laura express their skepticism about whether trees can survive the Plains climate. Almanzo notes that if trees could grow on the Plains, they would already be there, while Laura can't imagine the prairies covered in woods like the Big Woods of Wisconsin in her childhood. As it turns out, though Almanzo has the best of intentions in his desire to keep his trees alive, it is not quite as simple as merely planting the trees and watering them. Laura and Almanzo are justified in their skepticism.

Human activity did indeed change the weather, but not as the writers and railroad companies billed. The local climate changed as humans imposed their agricultural will on the land, especially with the increase in widespread irrigation. The weather patterns—the climate—changed more distinctly and more widely as humans changed the chemical composition of the atmosphere, through all of the activities supported by burning fossil fuels. Humans did not, however, impose their will on the Great Plains land to change its climate into a wetter one that would support sustained small-farm living. Drought is a natural part of the Plains climate, just as grasses are a natural cover to the prairies.

Drought Ends the Boom

As railroads extended into Dakota Territory from the late 1870s to mid-1880s and the Homestead Act provided land, the settler population surged in a period dubbed the "Great Dakota Boom." The population increased across Dakota Territory from just over 98,000 in 1880 to 348,600 in 1890.[19] Economic fortunes were tied closely

17. Jim McKee, "The Timber Culture Act: A Good Idea, A Dismal Failure," *Lincoln Journal-Star,* 17 July 2011, https://journalstar.com/news/local/jim-mckee-the-timber-culture-act-a-good-idea-a/article_ef8e6df2-b6df-57d7-81f7-9050db8c7fb8.html.

18. Wilder, *These Happy Golden Years,* p. 171.

19. South Dakota Department of Labor and Regulation, "South Dakota

to the success of agriculture and ranching, from the growing stages using the Homestead Act plots of land to the shipment of goods via the railroad network. Little did the settlers know how good they had it, how flush with rain they were. Homesteaders had unknowingly arrived on the Plains during a relatively wet (or "pluvial") period. Rain had not followed the plow; the plow had followed the rain. Homesteaders plowed up the prairie and planted crops with the mistaken impression that enough rain would continue to fall to support the acreage they tilled.

Then the pendulum swung, passing right through the "normal" middle-of-the-road rainfall and far into the dry side. To the homesteaders, the lack of rain must have felt as jarring as shifting from fifth gear to first in an old pickup truck. They were new to the Dakota climate still, not yet aware of how wide the swings could be between wet periods and dry. The skies and soils dried out just as the newly married Wilders began their farming years in the mid- to late 1880s, adding layers of difficulty from the very beginning of their years together. Laura had promised Almanzo just three years to give farming a try, and the drought sunk in its teeth by the end of their third year of marriage in 1888.

As in the 1870s during Wilder's Plum Creek days, a La Niña pattern established itself in the distant equatorial Pacific Ocean in the late 1880s through the late 1890s, dominating global weather patterns for nearly a decade. Recall that in the United States, La Niña patterns favor dry conditions in the southern Great Plains, often spreading northward into the central Great Plains when La Niña continues for multiple years. La Niña conditions occurred in 1886–1887, 1889–1890, and 1892–1895.[20] The La Niñas of the 1890s were

---

e-labor Bulletin" (August 2014), accessed 27 Mar. 2024, https://dlr.sd.gov/lmic/elaborbulletin.aspx; and U.S. Census Bureau, "Population of South Dakota by Counties and Minor Civil Divisions," *Census Bulletin* 47, 1 Feb. 1901, https://www2.census.gov/library/publications/decennial/1900/bulletins/demographic/47-population-sd.pdf. Note: One must add the Census population of 238,808 to the count of 109,792 "Indians and other persons" to reach the total population of 348,600.

20. McNoldy, "Historic Oceanic Niño Index and Niño 3.4 Data," accessed 8 May 2025, http://bmcnoldy.rsmas.miami.edu/tropics/oni/.

stronger and more consecutive than those of the 1870s, which can strengthen the downstream impacts. While not the only contributing factor, the persistent La Niña-like pattern was a significant driver of the dry weather patterns affecting wide swaths of the central United States.

In Huron, South Dakota, the driest summer on record (stretching back to 1881) remains the summer of 1894. The summer of 1893 ranks as eleventh driest and 1889 as seventeenth driest. In Huron, the dryness of the late 1880s through the mid-1890s was just about as profound as the 1930s "Dust Bowl" years. The temperatures for all three of those years (1889, 1893, and 1894) were above average, too, compared to an 1881–1910 average. In fact, the summertime average temperature in 1894 was the warmest on record until 1921, on level with the Dust Bowl years of the 1930s, and still ranks well above normal even including the warmer recent decades in the full record. Put a different way, the heat of the late 1880s to mid-1890s was roughly a once-in-a-generation event at the time, though it has become more of an every-other-year occasion in our modern climate.

The dry summers of the late 1880s into the 1890s were too much to take for many would-be farmers, and the settlers who were still so new to the land abandoned it in droves. The bust years peaked in the early to mid-1890s as relentless drought conditions caused crop failures year after year. Combined with factors such as lack of federal and state support for government relief for the farmers and a broader economic depression in the United States, drought was key to ending the boom of population expansion in the Dakotas.[21] Emigration from the Dakotas in the early 1890s caused a net loss in population from 348,600 in 1890 to 331,000 by 1895.[22]

The Wilders and their young daughter, Rose, were one of the families driven from their homestead by drought. Leaving De Smet

21 Richard Maxwell Brown, "The Enduring Frontier: The Impact of Weather on South Dakota History and Literature," *South Dakota History* 15, no. 1 (1985): 26–57; Herbert S. Schell, "Drought and Agriculture in Eastern South Dakota during the Eighteen Nineties," *Agricultural History* 5, no. 4 (1931): 162–80.

22. Allyson Brooks and Steph Jacon, *Homesteading and Agricultural Development Context* (South Dakota State Historical Preservation Center, 1994), https://history.sd.gov/preservation/docs/HomesteadAgDevelop.pdf.

in 1890, they moved in with Almanzo Wilder's parents in Spring Valley, Minnesota, for a year to find respite, and then to Westville, Florida, in search of opportunity. Disappointed there, the family returned to De Smet in 1892. They lived in town as Laura worked as a seamstress and Almanzo worked varying jobs, saving money so the family could resettle for good. The Wilders made no attempt to farm when they returned to De Smet, paralleling the Ingalls family experience of Wilder's childhood, in which they lived in town and did not farm when they returned to Walnut Grove after their unsuccessful year in Burr Oak, Iowa.

Charmed by a pamphlet advertising life in the Ozarks, the Wilder family made the decision to move to Missouri in 1894, at the climax of the drought. In an introduction to her mother's published travel journal, *On the Way Home*, Rose Wilder Lane later penned the conditions she had observed when they departed De Smet for good: "For seven years there had been too little rain. The prairies were dust. Day after day, summer after summer, the scorching winds blew the dust and the sun was brassy in a yellow sky. Crop after crop failed. Again and again the barren land had to be mortgaged, for taxes and food and next year's seed. The agony of hope ended when there was no harvest and no more credit, no money to pay interest and taxes; the banker took the land. Then the bank failed."[23]

The Dust Bowl Years

More of a regional aberration than a global one, the Dust Bowl of the 1930s was marked by several dry and hot years in the central and southern Great Plains, especially in a bullseye along the High Plains of western Texas, western Oklahoma, eastern Colorado, western Kansas, eastern New Mexico, western Nebraska, and western South Dakota.[24] The Dust Bowl displaced an estimated 2.5 mil-

23. Rose Wilder Lane, "On the Way Home: The Diary of a Trip from South Dakota to Mansfield, Missouri, in 1894," in *A Little House Traveler: Writings from Laura Ingalls Wilder's Journeys Across America*, ed. William T. Anderson (New York: Harper Collins, 2006), p. 5.

24. Defining the boundaries of the Dust Bowl area is not simple. For example, *see* https://courses.bowdoin.edu/history-2203-fall-2020-ialexand/your-choosing-i/, which shows how different definitions provide different geographical areas of impact.

lion farmers, ranchers, and other inhabitants of the Plains. Human activities like agriculture can weave together with weather and climate to create disaster if care is not taken to conserve the soil and the moisture resources beneath it. Chasing higher profits in an era of high supply, farmers plowed ever larger tracts of land to plant more quantity. Without the protective cover of native grasses, and without enough precipitation to support the kind of agriculture that had been introduced to the region in wetter years, the soil dried out and lofted in the hot winds, creating blinding, choking dust storms. The feedback process of hot, dry, vegetation-free ground cover exacerbated the heat, which worsened the evaporation out of the ground and dried the land even further.[25]

The result was a cascading series of devastating ecological, economic, and human health impacts. The drifted dust smothered the land, further preventing plant growth and blanketing the inhabitants and their dwellings. Residents of the region stayed in conditions that sickened and starved them, or they uprooted into an uncertain migration toward any greener pastures.

On the fringes of the Dust Bowl region, Missouri suffered from some drought in the 1930s, though not as extreme as in the core Dust Bowl region. Many of the migrants who plodded westward to California for new opportunities came from Missouri. The 1930s produced the third and ninth driest years on record in Springfield, Missouri: 1936 and 1930, respectively. The 1950s were even drier, with 1953 as the driest year on record and four of the ten driest years on record occurring between 1952 and 1956. The warmest year on record in Springfield remains 1938—by just a tenth of a degree—with three other years in the 1930s ranked among the ten warmest. The second, third, and fourth warmest years occurred in the 2010s, and the warmest twenty-five years are dominated by the 2000s and 2010s. What used to seem like a once-in-a-lifetime heat wave is becoming part of the new normal of temperatures in the region.

25. Timothy Egan, *The Worst Hard Time: The Untold Story of Those Who Survived the Great American Dust Bowl* (New York: Mariner Books, 2006), p. 340. This work provides the factual reference material for the Dust Bowl statistics cited in this paragraph.

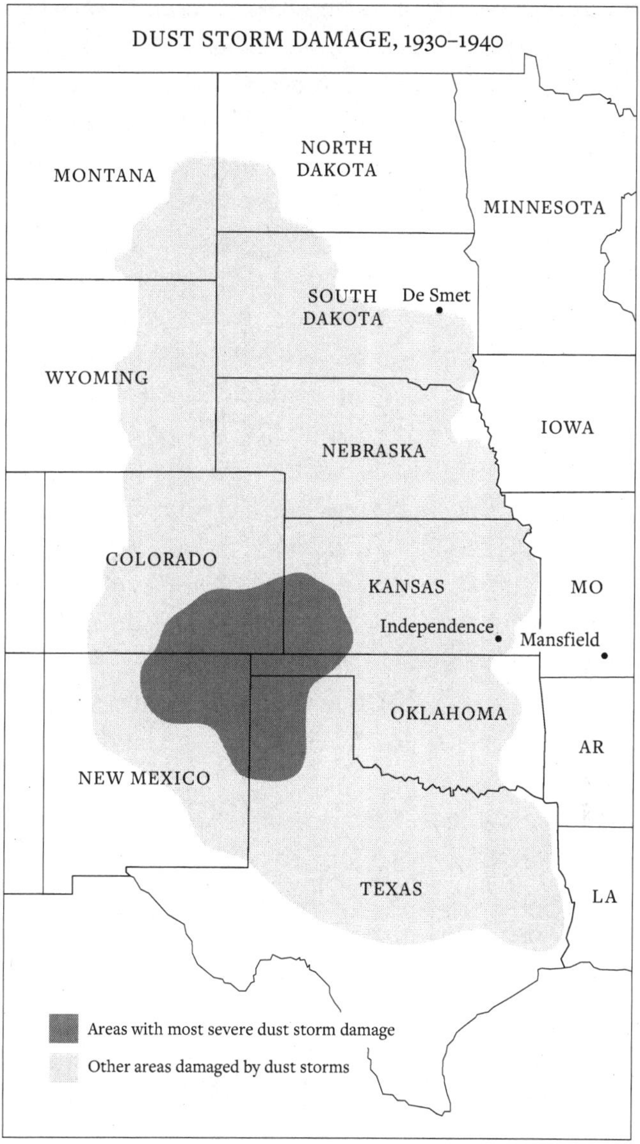

Defining the boundaries of the area impacted by the Dust Bowl is not simple. Different parts of the central U.S. were impacted in different years, some nearly every year for a decade and others only for a couple years. The heart of the Dust Bowl centered on the Texas and Oklahoma panhandles, northeastern New Mexico, southeastern Colorado, and southwestern Kansas, where repeated drought years and harmful farming practices combined to strip the land of its soil.

Southern Missouri did occasionally collect dust and dirt carried on the winds from the High Plains. Tucked into the Ozarks, Laura and Almanzo Wilder experienced occasional dust storms from the Plains to their west. For instance, Wilder wrote to Lane on 12 March 1937, "Dust settled over us again yesterday." They escaped the worst of the drought, however, as well as the most extreme heat.[26]

Winters during the Dust Bowl years were not as unusually warm as the summers, and like many in the Plains, the Wilders experienced wintry enough conditions during the 1930s to "write home" about to their daughter.

On 25 January 1938, Wilder wrote, "The creek rose up and roared! It was a warm rain soaking into the ground. Sunday night it turned cold and all night and all day and night Monday the wind blew from the southwest like the norther we met in Texas. I never before saw a wind like it here. It would freeze the marrow in one's bones. That night the wind stopped and next morning it was only 4 above zero. It is still cold."[27] On 28 January she added, "It is a beautiful, sunny morning but cold."[28]

A month earlier, on 29 December 1937 Wilder wrote to Lane, "We have had only a couple of days of sunshine for three weeks now. Fog so thick it is dangerous to drive even with lights on for they don't penetrate it. . . . There is no moisture and it is awfully dry here. Have you seen that the Mississippi River is lower than at any time since weather records have been kept in St. Louis in 1861?"[29]

As the dry 1930s continued, the Mississippi River indeed experienced low flows. The decade produced some of the lowest-ranked water levels on the Mississippi River bordering Missouri since written records began. The river has been engineered with locks and dams since then, which changes the possible range of river levels on many stretches of the river. Even so, river gauges along the Mississippi report several years in the 1930s with very low river levels,

26. LIW to RWL, 12 Mar. 1937, in Anderson, ed., *Selected Letters*, p. 114.

27. LIW to RWL, 25 Jan. 1938, in Anderson, ed., *Selected Letters*, p. 150.

28. LIW to RWL, 28 Jan. 1938, in Anderson, ed., *Selected Letters*, p. 154.

29. LIW to RWL, 29 Dec. 1937 (pp. 4–5 of that letter), in Anderson, ed., *Selected Letters*, p. 139.

including 1933, 1936, 1937, 1938, and 1940.[30] The river level on the Mississippi at St. Louis on 12 December 1937 remains the fourth lowest on record, and all the other years among the lowest water levels have occurred since 1940. When Wilder wrote her letter in December 1937, it appears she relayed accurate information about the record-setting low water flows; it had been the lowest flow ever recorded at the time.

On 6 January 1938, Wilder wrote to Lane, "We have been having nice weather lately, but dry. Letter from Grace today [from Manchester, South Dakota] says they are having a nice winter. That means a dry season next summer."[31]

Back in South Dakota, the farmers around the De Smet area did not fare well. Although South Dakota lay outside of the infamous Dust Bowl region of the southern High Plains, a second and less famous bullseye of dust and drought centered on central South Dakota to central North Dakota during those years.[32] The worst of the dust and soil erosion in central South Dakota clipped Kingsbury County, of which De Smet is the county seat, creating hardships on par with those in the southern Plains. Plants withered in the drought and heat. Topsoil turned to dust, whipped into black blizzards by the winds. Food and water were scarce, hunger abundant. Farmers and ranchers gave up their lands and operations, climbed into rickety cars overloaded with their few possessions, and sought new opportunities. As with every state from North Dakota to Texas, South Dakota's population dwindled, from 692,848 in the 1930 census to 642,961 in 1940; it did not return to 1930 levels until the 1980s.[33]

30. River gauge records are available site by site via the National Weather Service's river hydrology website, https://water.weather.gov/ahps/.

31. LIW to RWL, 6 Jan. 1938 (pp. 4–5 of that letter), in Anderson, ed., *Selected Letters*, p. 144.

32. Taylor L. Weeden, "Was There a Northern Dust Bowl? Evidence for Heightened Wind Erosion and Dust Sources during the 1930s in the Northern Great Plains, USA" (MS diss., Baylor University, 2021).

33. "Historical Population Change Data (1910–2020)," U.S. Census Bureau, 26 Apr. 2021, https://www.census.gov/data/tables/time-series/dec/popchange-data-text.html.

In a 1994 interview for South Dakota Public Radio, Roy and Lucille Fitz shared their own experiences as a young family in De Smet during the Dust Bowl years. "Drouth," as the old-timers still sometimes say, settled into the De Smet area as early as 1929, with conditions growing hotter and drier each year through 1933. Six weeks of hot winds in 1930 cooked the flax crop in bloom, destroying its pollen and stifling flaxseed growth. Bad as that was, 1931 was "much worse," with the arrival of dust storms and tumbleweeds (also known as Russian thistles). Roy remembered 11 November 1933 as a "black blizzard." By 1933, many farmers had thinned their herds or lost their land. According to Roy, statistics indicate that only 39 percent of farmers lost their land, but to him, in the De Smet area, "That was absolutely wrong. I would say it was eighty-nine percent. My dad nearly lost his land."[34]

After a mildly successful crop year in 1935, "if you had the courage and the credit," as Roy put it, the hottest year of all arrived in 1936. July 1936 remains on the record books with the hottest temperatures ever recorded in many central and northern Great Plains cities. Roy remembers the temperature reaching 115°F in De Smet; official records from July 1936 list the hottest maximum temperatures on record: Mobridge, South Dakota (116°F), Norfolk, Nebraska (116), Concordia, Kansas (116), Aberdeen, South Dakota (115), Lincoln, Nebraska (115), Fargo, North Dakota (114), Bismarck, North Dakota (114), Omaha, Nebraska (114), Topeka, Kansas (114), Sisseton, South Dakota (113), Kansas City, Missouri (113), Sioux City, Iowa (111), Sioux Falls, South Dakota (110), Des Moines, Iowa (110), Grand Forks, North Dakota (109), Minneapolis/St. Paul, Minnesota (108), and many more.

The Dust Bowl conditions prolonged the Great Depression in the central United States. Land management practices helped soil conditions improve later in the decade, but economic conditions only recovered in the early 1940s, mostly due to a combination of weather and war, with average to above-average rains returning to the Plains and with the economic benefits of World War II.

34. Roy and Lucille Fitz, interview by Chuck Anderson, *Personalities with Chuck Anderson*, South Dakota Public Radio, 22 Oct. 1994, https://listen.sdpb.org/arts-culture/2016-09-19/personalities-roy-and-lucille-fitz-pt-ii.

Drought is a great integrator of weather, climate, and environment. It is often fed by seasonal weather patterns or long-term changes. Frequently introduced or exacerbated by short-term weather patterns, like hot and dry spells, hot and windy weather, or even the lack of other weather hazards like thunderstorms and hurricanes, the effects of a drought can be both improved and worsened by human interventions like irrigation, reservoir management, or the presence or lack of community planning. Drought can cure fuels into tinder for wildfire, dry up the creeks and ponds needed for many species of wildlife, and create expensive gaps in income for farmers, ranchers, and those who depend on tourism and recreation.

Among its many ills, drought can also bring pests. Bark beetles feast on drying and dying pine trees in Colorado. Tiny spider mites, as insipid as the drought itself, thrive in hot and dry weather. Ticks and spiders seek moisture inside homes. Trees, already under stress from drought, are targeted by outbreaks of plant-eating insects and fungi who seek their concentrated sugars.

The contributors to and victims of drought are entangled. To be sure, weather and climate lay the foundations for drought. Drought affects the entire ecosystem: the waterways, soils, plants, and animals living within the drought's fingerprint. Humans shape the drought in the way we manage the land, use the waters, and grow the plants. Humans can build reservoirs that hold melt waters from mountain snowpack so there is a pool for the dry summer months, easing the sting of a drought. As in the Dust Bowl years, they can till and plow the land without enough water to support the crops they are trying to grow, leaving the naked soil prone to evaporation and desiccation and turning it to dust. Though it's sometimes uncomfortable to think about, in a drought, when water is scarce, the choice to wash a car or water a lawn might mean not leaving enough to keep a trout stream flowing or irrigate a local farm.

As with all weather and climate, humans can both worsen a natural disaster and be a victim of it. We build in fire- and flood-prone boundary lands, water our lawns during drought, and—yes—drive our cars and fly on planes. It doesn't make what we do right or wrong—often, it's the only choice we have, or it's the best choice among others when weighing all the factors that are important to us. Collectively, though, our individual and community choices do have an impact. We humans are not separate from the world around us. The weather and climate put their marks on us, and in many ways, we put our marks on them, too.

While technology has advanced substantially in weather observations and forecasting since the U.S. government began weather operations in 1870, some instruments are still the same. On the right, a young observer (or, more likely, the child of an observer) stands next to the "cotton region shelter" that houses thermometers to measure both temperature and wet bulb temperature (which approximates to humidity). On the left, the observer stands next to the rain gauge at the observing site. Both photos are taken at a Cooperative Observer Program site, circa 1930s, in Granger, Utah. *Courtesy National Oceanic and Atmospheric Administration*

On the left, a shelter houses weather observation instruments with a rain gauge in the foreground. The photo at the bottom provides a closer view of the two thermometers inside the shelter, one to measure temperature and the other on a spinner to measure wet bulb temperature. Both photos were taken at a Cooperative Observer Program station near Houghton Lake, Michigan, in 2021 by the author.

This 1894 portrait depicts the Ingalls family shortly before Laura and Almanzo Wilder moved to Mansfield, Missouri. Seated, from left to right: Caroline, Charles, and Mary Ingalls. Standing: Carrie Ingalls, Laura Ingalls Wilder, and Grace Ingalls. *South Dakota State Historical Society*

This US Wildlife Service photo shows a typical muskrat mound, providing a convenient resting spot for two Canada geese near Huron, South Dakota. *Courtesy USFWS Mountain-Prairie Region's Flickr photostream by photographer Sandra Uecker*

This wintry view of De Smet, South Dakota, includes the Ingalls building.
South Dakota State Historical Society

Bird's-eye view of De Smet, 1883. *South Dakota State Historical Society*

Photographer George I. Vance captured these images of snowbound trains along the Winona & St. Peter Railroad in Minnesota, providing a glimpse of the herculean challenge facing the men tasked with clearing the tracks. *Courtesy Chicago & North Western Railway Historical Society*

Laura Ingalls Wilder and Almanzo Wilder posed for this photograph soon after their wedding in 1885. *Courtesy Laura Ingalls Wilder Home Association, Mansfield, Mo.*

Laura Ingalls Wilder wrote her Little House books at the farmhouse she and Almanzo built out of materials found on their Rocky Ridge Farm. *South Dakota State Historical Society*

Laura Ingalls Wilder was twenty-seven years old when she posed for this portrait.
South Dakota State Historical Society

Rose Wilder Lane posed for this portrait in the late 1920s or early 1930s.
Courtesy Herbert Hoover Presidential Library

Drifting dust buries farm machinery near Gregory, South Dakota, at the home of County Commissioner C. M. Olsen in 1936. *South Dakota State Historical Society*

Floodwaters inundated part of Yankton (top), Dakota Territory, in late March 1881. An ice jam flood caused the damage seen in this photo of Vermillion (bottom), Dakota Territory, also in late March 1881. *South Dakota State Historical Society*

(top) Terrible fires raged in the Midwest on 8–10 October 1871. The Great Chicago Fire destroyed much of the city, including 17,450 buildings, claimed 300 lives, and left 100,000 people homeless. Laura's De Smet friend Ida Brown lost her parents in the fire. *Courtesy Library of Congress.*

(bottom) The Peshtigo, Wisconsin, fire burning at the same time as the Chicago fire caused even more damage, burning 1,875 square miles, destroying twelve communities, and claiming between 1,200 and 2,500 lives. The Peshtigo Fire is considered the most devastating forest fire in U.S. History. This *Harper's Weekly* engraving shows the desperation of people seeking refuge from the fire in the Peshtigo River. *Courtesy Wisconsin Historical Society*

Dead grasshoppers are piled in a heap in Pierre on 1 August 1933 (left).
South Dakota State Historical Society

Grasshoppers climb the wall of a Pierre home on 1 August 1933 (right). Note that the grasshoppers have eaten some paint from the house and leaves and bark from the trees.
South Dakota State Historical Society

Rose Wilder Lane took this photo of Laura Ingalls Wilder, age eighty-seven, in the front yard at Rocky Ridge in March 1954. *Courtesy of Laura Ingalls Wilder Home Association, Mansfield, Mo.*

III

Connect Weather and Climate to People and Their Environments

9

Wilder Fires

Few locations in the United States are truly insulated from the risk of wildfire, though some places carry more risk than others. Perhaps because fires never directly threatened my safety, my memories of fires through my life are vague. I do remember the Stephan Bridge Road Fire, which ignited north of Roscommon, Michigan, the nearest town to my childhood home, in early May 1990.[1] Unlike Laura Ingalls Wilder's sharp recollection of the Hard Winter a half-century after the fact, the only thing I remember about the weather at the time was that it was dry, and I had to look it up to confirm what year the fire occurred.

April 1990 was much drier and much warmer than normal at Grayling, Michigan, the nearest COOP station to the Stephan Bridge Road Fire. The first half of the month was on the cool and wet side, but beginning

1. National Fire Protection Association, "Stephan Bridge Road Fire Case Study," March 2018, accessed 10 Apr. 2024, https://www.denix.osd.mil/dodpif/denix-files/sites/37/2018/03/1990-National-Fire-Protection-Association.-Stephan-Bridge-Road-Fire-Case-Study.pdf; "80 homes are destroyed by wildfires in Michigan," *New York Times*, 10 May 1990, https://www.nytimes.com/1990/05/10/us/80-homes-are-destroyed-by-wildfires-in-michigan.html; and UPI, "Forest fires ravage northern Michigan," 9 May 1990, https://www.upi.com/Archives/1990/05/09/Forest-fires-ravage-northern-Michigan/8505642225600/.

on 19 April, temperatures quickly turned very warm. Two days in April set daily maximum temperature records that still stand in 2024. The nearest first-order station, in Houghton Lake, Michigan, tells the same story: hot and dry weather from the end of April into early May. Precipitation ceased after 20 April, not to return until 4 May—one of just three years on record with no measurable precipitation through those dates. Even through 2024, the last two weeks of April 1990 stand in the records as the fifth warmest and thirteenth driest.

A couple of days in early May flinched from the hot and dry streak, with cooler temperatures and even a little rain. It wasn't enough. The fire ignited on 8 May, after two more hot and dry days.

Even now, I scold myself that I can't reliably remember how it felt to experience upper-eighties temperatures in northern Michigan in late April. I probably wore a skirt to school and enjoyed not shivering as I waited for the bus at seven in the morning. Maybe I rode my bike after school, or maybe I hiked into the adjacent North Higgins Lake State Park. Maybe I joked about going for a swim in the ice-cold waters of Higgins Lake, visible from my front yard.

Perhaps it was just by the power of suggestion of seeing the towering plume, but I remember that the air smelled thick with smoke and that ash fell like snowflakes. Because I lived next to a state park, the smell of a smoky campfire infuses my summer memories, but the wildfire smoke smelled more intense, more oily, and it clung to its surroundings like fog to a coastline. From the middle school playground, located in downtown Roscommon, we surely could see the smoke, and maybe that's where I remember ash. But my most vivid memory of the plume is seeing it from the roads near the north side of Higgins Lake, farther away from the fire than downtown Roscommon.

The fire's distance and direction of movement meant our town and my home were never in its path. Even so, I had friends who lived on and near Stephan Bridge Road. The forests between that wilderness north of Roscommon and east of Grayling and my home lay more or less continuous, much of it in protected swaths of state forest. The roads cutting through the forest were hardly wide enough to part the canopy, let alone provide a firebreak. Only the wider I-75 corridor offered a boundary, but I knew fires could jump the interstates in "up north" Michigan; the two-lane county roads that lay between the fire and downtown Roscommon would have been no better protection than the forest roads. Had the winds been out of the northeast instead of the northwest, nothing would have stopped the fire from jumping tree to tree, from combustible jack pines to white pines to stout oaks, and into town. The notion raised the hair on my preadolescent arms, even if my house would have been at less risk than the town itself.

The expansive, cool waters of Higgins Lake, I always knew, were my safe haven in a fire. If one of the state park campfires ever got out of hand, if fire ignited in the woods between towns and rode the wind toward us, I knew I could trust Higgins Lake to keep me safe. The dash to the beach was only a few hundred feet. I could wade out to the drop-off and duck my head underwater as the ashes flew. My family could lower our boat from its hoist, start its motor, and speed to the middle of the large lake, far from the singeing heat and embers.

Fires seemed to come even more frequently after I moved away from northern Michigan. The Four Mile Fire in April 2008 burned much closer to my home, between Higgins Lake and Grayling.[2] Residents evacuated their homes. I-75 closed as fire encroached and jumped the road, the lone limited-access highway going north-to-south from the Mackinac Bridge to the interstates that connected our world to the big cities of southern Michigan. The fire threatened Grayling itself and the gas stations on the south side of town, adjacent to the tree line, with only the clearing space of the interstate's exit ramp as a barrier.

The Meridian Boundary Fire burned in mid-May 2010 northeast of Roscommon, within a few miles of the Four Mile Fire's burn swath, initiated when a (legal) leaf burn got out of control.[3] On the same day, the Range 9 Fire ignited near Camp Grayling, a Michi-

2. Amy L. Payne, "Blaze that Scorched 1,100 Acres, Skirted Grayling Brought Under Control," *MLive*, 25 Apr. 2008, https://www.mlive.com/bctimes/2008/04/fire_burns_through_1100_acres_1.html.

3. Bill Gabbert, "Update on the Michigan Fires, May 19," *Wildfire Today*, 19 May 2010, https://wildfiretoday.com/2010/05/19/update-on-the-michigan-fires

gan National Guard training ground southwest of Grayling—again, closer to my childhood home. Yet another, the Wilderness Trail Fire, burned in early June 2023, sparked by a campfire and fueled by a dried-out forest.[4] This fire, like the Four Mile Fire, started southeast of Grayling, but it burned closer to town and threatened (and closed) I-75.

As each burn scar has healed, another has lined up to replace it, almost on cue. It isn't lost on me that the fires seem to be coming closer and closer together, though I don't have enough information to say if it is a trend, a change in fire management practices, or an unlucky coincidence.

Fire weather still raises the hairs on my arms in the grassland prairies of eastern Nebraska. Early spring in a dry year and fall of just about any year carries fire risks for the same reasons across the seasons: crunchy vegetation, gusty winds, and dry air masses. The warm winds and brown, winter-cured vegetation on a fire weather day are no match for sparks. I live in suburbia now, but grasslands and farm fields still edge up to the boundaries of our neighborhood.

A farmer will put off taking his tractor down the road on fire weather days, because it has a dangling chain that can sometimes

---

-may-19/; Bill Gabbert, "Map of Meridian Fire in Northern Michigan," *Wildfire Today*, 19 May 2010, https://wildfiretoday.com/2010/05/19/map-of-meridian-fire-in-northern-michigan/; and Edward Vielmetti, "Grayling Area Fires—Range 9 Fire, Meridian Boundary Fire," WordPress, 20 May 2010, https://vielmetti.wordpress.com/2010/05/20/grayling-area-fires-range-9-fire-meridian-boundary-fire/.

4. Associated Press, "Michigan Wildfire that's Burned More than 3 Square Miles Was Sparked by Campfire on Private Land," 4 June 2023, https://apnews.com/article/michigan-wildfire-hot-dry-weather-4f14187d85bdf76bfbd45ac8d657ed0d; Brandon Chew, "Wilderness Trail Fire 100% Contained, Michigan DNR Says," Michigan Department of Natural Resources, 6 June 2023, https://upnorthlive.com/news/local/wilderness-trail-fire-100-contained-michigan-dnr-says; "Wilderness Trail Fire Map Updated," *Traverse City Record-Eagle*, 4 June 2023, https://www.record-eagle.com/wilderness-trail-fire-map-update-pdf/pdf_638700b4-0309-11ee-be07-033cadf572a7.html; and Mike Horace, Teresa Homsi, Rick Brewer, and Amy Robinson, "Grayling Wildfire Caused by Campfire; DNR Says Fire More than 90% Contained," WCMU Public Media, 5 June 2023, https://radio.wcmu.org/2023-06-03/large-wildfire-reported-southeast-of-grayling.

throw a spark when it strikes pavement or gravel rocks. NWS safety messages urge drivers not to throw their cigarette butts out the window. Many highways are flanked with scorched grass, sometimes spilling into adjacent fields and shrubs, the scars of careless drivers who probably never knew that they touched a match to a powder keg.

By afternoon on a fire weather day, thin smoke plumes can dot the horizon. Our eyes scan for them and gauge wind direction. Is that plume upwind? Will this one stay on the other side of the interstate? Even weather tools like satellite and radar light up with fires, their plumes visible as specks of warmer temperatures through the lens of a satellite sensor, or as a stripe of particles that catch and reflect the weather radar beams as they loft and travel with the wind. At the peak of the afternoon heat and winds, fire pixels and plumes speckle the scope.

Most days, local firefighters tamp down the grass fires as a part of routine operations. Once in a while, one of those fires catches hold of cured fuels, or perhaps spreads into terrain that isn't accessible enough for firefighters to get ahead of it, and blossoms in size and intensity. Across much of the central Plains, including Nebraska, 2022 was a dry year. In April, a grassland wildfire ignited south of Elwood, Nebraska, and approached the town of Arapahoe. Elwood's Fire and Rescue chief died in a collision between his SUV and a water truck in zero-visibility conditions due to smoke.[5] In October, an ATV operating off trail ignited the Bovee Fire in western Nebraska near Halsey, which blossomed into a large fire that burned up a 4-H camp and a few farmsteads.[6] The volunteer as-

5. "Elwood (NE) Fire Chief Killed Fighting Wildfire, Died in Crash with Water Truck," Firefighter Nation, 8 Apr. 2022, https://www.firefighternation.com/news/elwood-ne-fire-chief-killed-fighting-wildfire-killed-in-crash-with-water-truck/#gref; and Mary Jane Skalka, "The late Elwood Fire Chief Darren Krull will be remembered at national memorial this weekend," *Kearney Hub*, 3 May 2023, https://kearneyhub.com/news/local/the-late-elwood-fire-chief-darren-krull-will-be-remembered-at-national-memorial-this-weekend/article_c2be7e8c-e9c2-11ed-a808-4f77ebc12bec.html.

6. Carson Vaughan, "Nebraska National Forest Bovee Fire Destroys Beloved 4-H Campground," High Plains Public Radio, 7 Oct. 2022, https://www

sistant fire chief of Purdum, Nebraska, died of a heart attack while fighting the fire.[7]

While fires are part of the normal landscape in many areas, their intensity, size, and frequency tangle with human influences such as fire management practices and building in prone locations, along with the changes to weather in some places that make hotter, drier weather more common. Even in ecosystems where fire is healthy, it can and does behave erratically and it can affect people in unhealthy and dangerous ways.

Forest Fires and the Great Fires of 1871

As bad as Plains grass fires can be, fires take on an apocalyptic level of heat and fright in the West. With miles of forests ablaze, entire towns evacuate in panic as fires advance as fast as a car can drive, treetops explode into flames, and fire-fueled winds whip the flames and embers ahead.

The smoke of large forest fires can carry for thousands of miles. From source fires in California, British Columbia, Colorado, and other hot spots made volatile by the weather and climate and dry fuels, smoke whitens the skies over places as far as Oklahoma, Illinois, or even the East Coast. Carried high in the atmosphere, the smoke can be indistinguishable from a cover of high-level clouds. Yet there is a faint pungent smell in the air, as though a neighbor some distance away has lit a bonfire; the air is not quite clean, and those who struggle to breathe may struggle a little harder.

Smoky skies make for hazy days but brilliant sunrises and sunsets. As the sun rises and sinks, its rays pass through the billions of particles of ash and soot, and the skies blaze red and orange, as though reflecting the fires themselves. Shades of blue melt away until the winds above change directions, or the fires are extinguished, and the smoke clears.

.hppr.org/hppr-news/2022-10-07/nebraska-national-forest-bovee-fire-destroys-beloved-4-h-campground.

7. "NE Assistant Chief Dies Battling Bovee Wildfire," Firehouse, 4 Oct. 2022, https://www.firehouse.com/lodds/news/21282716/purdum-ne-assistant-fire-chief-mike-moody-dies-fighting-wildfire.

Closer to the source, the smoke is more acrid, thicker, tangible. Traces of ash settle over everything. The smell permeates clothing and clings to skin and hair. Vulnerable people, those who are very young, very old, or already ill, can struggle to breathe. The smoky cloud may glow where it scrapes the ground, lit by the source fire underneath.

Wind steers a wildfire's direction and speed. Hot temperatures and low humidity can make fires both ignite more easily and travel more efficiently on the winds. Thus, the most volatile weather for wildfire activity is hot, dry, and windy.

Fuels on the ground stoke the fire with energy. Dried grasses burn swifter and with less heat than wood. Forest fires burn both the leaf and limb debris and underbrush on the forest floor, as well as in the canopy, when the tops of the trees themselves burst into flames. Canopy fires are especially hot and dangerous, engulfing the entire forest cover and leaping through treetops across firebreaks. Modern wildland management practice includes a mix of letting some fires burn and working quickly to extinguish others. Human lives and property must be protected, but fires are an important and natural part of forest and grassland ecosystems that clear away old growth so new growth can occur.[8]

Fires can even create their own weather. Their heat can generate gusty winds around them that further accelerate their movement. The hottest of fires can cause fire whirls or fire tornadoes, violently rotating columns of fire causing both burn and wind damage where they spin. The smoke above a hot fire can take on the shape and characteristics of a thunderstorm's towering cumulonimbus cloud. Called a "pyrocumulus" cloud, it can generate its own lightning and be visible for over a hundred miles distance from the source fire.

Humans and animals alike flee from advancing fires, often successfully but sometimes with mixed success. Weather and fire fore-

8. Craig A. Harper, W. Mark Ford, Marcus A. Lashley, Christopher E. Moorman, and Michael C. Stambaugh, "Fire Effects on Wildlife in the Central Hardwoods and Appalachian Regions," *Fire Ecology* 12 (2016): 127–59, https://doi.org/10.4996/fireecology.1202127; and Jacob Margolis, *The Big Burn* podcast: "Setting a Forest on Fire, LAist Studios/NPR, 12 Oct. 2022, accessed 13 Apr. 2024, https://www.npr.org/podcasts/674580962/the-big-disaster-the-big-burn.

casts often allow people enough notice to get themselves to safety, though their possessions are another matter. The worst wildfires in grasslands and forests alike can advance too quickly and burn too hot for humans or animals to escape.

Though forest fires are usually associated with the western United States, especially in recent decades, the Great Lakes region has also been vulnerable to them, the legacy of careless logging practices in the late 1800s that left the woodlands especially fire prone, with stumps and tree debris littering the forest floor.

The Ingalls family returned to the woods of western Wisconsin from their stay on the Kansas prairie in May 1871, a somewhat dry year in the Great Lakes. The total rainfall from September through mid-October of 1871 remains in the top ten driest at first-order stations around the upper Great Lakes region (Minneapolis/St. Paul, Madison, Milwaukee, Marquette, Lansing, and Chicago). The summer had been just on the dry side of average for the same region. The same "flash droughts" that bring dried-up crops to the Corn Belt can also stoke fires by abruptly curing the natural vegetation in a forest, especially the understory plants and shrubs.

Across Wisconsin and the Lower Peninsula of Michigan, the abruptly dry fall weather prompted the leaves to shrivel and crumple to the ground into a brown, crinkling mass, rather than remain as a glorious canopy of reds and yellows. The debris and stumps from careless logging baked in summer warmth and fall dryness, cured into tinder that was yearning for a spark. Sporadic fires burned through September and into October, with notable fires on 23–24 September in eastern Wisconsin. On 5 October, meteorologist and observer Increase Lapham, based in Milwaukee, noted, "Fires all the way from Lake Michigan to Dakota Territory."[9]

On 8 October, sparks fell, gusty winds whipped small fires into conflagrations, and flames leapt into the skies above the tree canopies of Wisconsin, Michigan, and Illinois. The source of the sparks is not known and is a source of great speculation, ranging from dry lightning to railroad crews and activity to comets and mete-

9. Denise Gess and William Lutz, *Firestorm at Peshtigo: A Town, Its People, and the Deadliest Fire in American History* (New York: Holt Paperback, 2002), p. 82.

ors (almost certainly not).[10] The most likely explanation is that the small fires that had been burning across the region since September converged and accelerated into much larger fires as strong winds blew into the region. Numerous great fires swept through tracts of land, especially in eastern Wisconsin and the southern half of lower Michigan.

The fire in Peshtigo, Wisconsin, the deadliest of the day, took as many as 2,500 lives in the town and the surrounding woods and communities. Originating near Green Bay, it blasted through Peshtigo and eventually reached Marinette, Wisconsin, and neighboring Menominee, Michigan. Fire also burned on the other side of Green Bay, into Door Peninsula. The fire burned around 1.5 million acres (2,400 square miles) of land, an area larger than the state of Delaware, and pushed residents against the banks of the Peshtigo and Menominee Rivers and the shores of Green Bay.[11] Many survivors reported fire-tornado-like conditions and wind speeds, with the fire-fueled winds anywhere from hurricane force to 100 mph or more.[12] The devastation left costly and generational economic impacts, stifling the local economies along with the influx of immigrants to the region.

On the same day, a series of wildfires collectively known as the Great Michigan Fire burned across the Lower Peninsula of Michigan.[13] Of the three primary swaths, one burned along the Lake Michigan coast, another up the thumb of the Lower Peninsula, and the third along the Au Sable River in northern lower Michigan.[14] Fires destroyed part or all of the towns of Port Huron, Holland, Manistee, Grayling, and Alpena. Survivors again described "fire tornado" conditions, with gusty winds already in place and

10. Ibid., pp. 86, 188–89; Brandon James, "Mysterious 'Meteor Theory' For Michigan's Great Fires Of 1871," WBCK, 8 Sept. 2018, https://wbckfm.com/michigan-great-fire-1871/.

11. Gess and Lutz, *Firestorm at Peshtigo*, p. 211.

12. Ibid., p. 211.

13. Ibid., pp. 113, 165, 169, 207.

14. Steward H. Holbrook, *Burning an Empire: The Story of American Forest Fires* (New York: Macmillan Company, 1943), p. 94, https://reader.library.cornell.edu/docviewer/digital?id=chla2900624#page/4/mode/1up.

fire-fueled winds whipping them to even greater speeds.[15] The fires across Michigan killed at least five hundred people, but the number of dead is not well known, as lumberjacks, remote settlers, and other people scattered across the state may not have been reported as missing.[16] The Great Michigan Fire burned an estimate of at least 2.5 million acres (3,900 square miles) across the state, or roughly the size of Delaware and Rhode Island combined. Put another way, the fires burned about 7 percent of Michigan's land area.[17]

The Peshtigo and Great Michigan Fires were overshadowed, however, by the much more famous Great Chicago Fire of 1871, which started on the same day and burned through 10 October. While the legend of Mrs. Kate O'Leary's cow lighting the fire by kicking over a lantern has not held up to scrutiny, the land around Chicago certainly was dry as tinder and prone to ignite, whatever the cause.[18] The flames swept through the city's closely packed wood-framed buildings like touching matches to newspaper, from the west side of the city straight through the heart of downtown Chicago. It dissipated only when it reached the north side of Chicago, where undeveloped land and the waters of Lake Michigan halted the fire's advance. The devastating fire caused about three hundred deaths and forced the city of Chicago to rebuild almost from scratch.[19]

Among the victims of the Great Chicago Fire of 1871 were the birth parents of Ida Wright Brown, Laura's dear friend in De Smet.[20] Ida was sent to an orphanage after the fire, where Reverend

15. Donald A. Haines and Rodney W. Sando, *Climatic Conditions Preceding Historically Great Fires in the North Central Region* (U.S.D.A. Forest Service Research Paper NC-34, 1969), https://web.archive.org/web/20120208132439/http://www.ncrs.fs.fed.us/pubs/rp/rp_nc034.pdf.

16. Ibid.

17. Greg Tasker, "Histories and Mysteries: The Great Michigan Fire," *Northern Express*, 18 Mar. 2023, https://www.northernexpress.com/news/feature/the-great-michigan-fire-1871-Manistee/.

18. Holbrook, *Burning an Empire*, p. 101.

19. *Exoneration of Mrs. O'Leary and Her Cow from All Blame in Regard to Great Chicago Fire of 1871* (Chicago City Council, 1997), https://www.chipublib.org/wp-content/uploads/sites/3/2014/09/counciljournal10281997.pdf.

20. Jennifer Gonzales and Annie Ross, "The Great Chicago Fire," *Library of Congress Blog*, 3 June 2022, https://blogs.loc.gov/law/2022/06/the-great-chicago-fire/.

and Mrs. Brown adopted her. They took their adopted child west as the minister expanded his missionary field into Dakota Territory, where Laura and Ida met in school. Ida and her fiancé, Elmer, served as the witnesses to Laura and Almanzo Wilder's wedding, with Reverend Brown presiding.[21]

In her *Pioneer Girl* manuscript, Wilder took note of the fires: "One day the sun was nearly hidden by smoke all day and when dark came the sky was reddened by fire. We stood in the door watching it and soon we could see fire run up to the tops of some trees on a hill and then the trees stood there burning like great candles."[22] She was four in the fall of 1871, old enough for the smell of smoke to be imprinted in her memory. She and her family were fortunate to watch the fires only from afar; many other settlers were not so lucky.

Kansas Prairie Fire

Every spring, the Flint Hills of Kansas burn.

Encompassing a hilly, rocky, sparsely settled north-to-south strip of land east of I-135 from Salina to Wichita and west of the interstates and highways from Topeka to Emporia to Eureka, the Flint Hills are blanketed with the last stand of original tallgrass prairie in the United States. Every spring, the air dries, the snow (if any) melts, and ranchers set intentional fires, a practice called "prescribed burning." A string of glowing orange specks dots the hills, flickering like candles in the distance, visible to passing cars. The constellation of fires shifts shapes through the spring as the fires singe grasslands from brown to black, soon to turn green again. Many days, the south to southwest winds that dominate in spring loft the smoke across eastern Kansas into adjacent Nebraska and Missouri, sending residents to their doorsteps with wrinkled noses to look for the source of the leafy, subtle smoke. The southern skies in Nebraska fade into a milky white haze, while the air in eastern Kansas is thick with smoke, dropping lines of sight to just a few miles. Eastern Kansans are used to coughing on thick smoke in the spring, but many city dwellers in Lincoln and Omaha forget that the

21. Wilder, *Pioneer Girl: The Annotated Autobiography*, p. 244.

22. Ibid., p. 322; *De Smet Leader*, 25 Aug. 1885.

Kansas smoke will carry with strong south winds. They cough, too, and call their weather broadcasters and officials to ask about the source and post their opposition to Kansas smoke on social media.

In many ecologies, periodic wildfires are an important part of the health of the system. They remove scrub and invasive vegetation, spread seeds, and enrich the soils.[23] In the modern Flint Hills of Kansas, ranch owners burn the tallgrass prairie annually to encourage fresh grass growth and subdue shrub growth. Independence, Kansas, the nearest town to the *Little House on the Prairie* homestead, lies one county east of the Flint Hills, or about twenty-five miles outside of them. There, too, the tallgrass prairie of old and the prairies of today were made to burn. The burning of the grasses is optimal in spring, while vegetation is still browned from winter but after any snow cover has melted and any lingering dampness has evaporated away. Fresh growth emerging with spring warming will push through the charred soil more easily than it would through a tangle of old, dead growth.

American Indians who lived in the Flint Hills and nearby plains had a long history of setting intentional fires, for the same reasons that non-native settlers did later. The Kansa (Kaw), Pawnee, Wichita, and Osage peoples observed how the grasses grew back greener and more tender after lightning-induced fires, thereby attracting bison, deer, and other grazing animals. As a part of their cultural practice of tending to the land, to induce fresh growth, clear shrubs, and support their hunting practices, these tribes and their ancestors had set intentional burns for centuries before European-descended explorers and settlers reached the Flint Hills, and continued these practices right up until they were forcibly removed from the land.[24]

Given the right combination of weather and fuel conditions, prescribed burns can grow out of control in a matter of minutes. Ideally, for safe controlled burns, winds should be something more than calm but something less than gusty so that the fire can spread

23. Jenn Little, "EPA Region 7 Research Spotlight: Studying Fire in the Flint Hills," EPA in Kansas, accessed 17 Apr. 2024, https://www.epa.gov/ks/epa-region-7-research-spotlight-studying-fire-flint-hills.

24. "Flint Hills First Peoples," National Park Service, accessed 17 Apr. 2024, https://www.nps.gov/articles/000/flint-hills-first-peoples.htm.

and smoke can clear from overhead, while not racing ahead of the fire's managers or across human-made or natural firebreaks such as trenches and creeks. Grassland fires can become dangerous when winds exceed around 15 to 25 mph, especially when the relative humidity is around 20 to 25 percent or lower and fuels are dry. The combination of those factors in the modern weather world sparks a red flag warning, a notice from the NWS that fires could become uncontrollable quickly. County officials may institute burn bans under the dangerous conditions, encouraging anyone with plans to set a controlled fire to wait until conditions are less volatile.

In *Little House on the Prairie*, on a spring day when Pa is plowing his field and Ma and the girls are at the house, Ma spies a dark cloud in the southern sky and wonders if a storm is brewing. Pa races in from the field a moment later, alerting the family to the prairie fire creating the cloud. He shouts instructions to the family to get a tub of water and soak sacks in it. Rabbits, deer, and other animals race across their perimeter ahead of the flames, bounding past their house and toward the creek. As the children aid Ma, Pa plows a furrow behind their horses, Pet and Patty, to act as a firebreak. He only has time to plow once around the house: "That fire's coming faster than a horse can run," he exclaims.[25] The furrow will not be wide enough to stop the wind-driven flames.

Winds of 25 to 30 mph are relatively common during the spring in eastern Kansas.[26] Gusty southwest winds in the spring would bring warmer temperatures, and they often are dry winds. Northwest winds can also be dry and warm enough in spring to support red flag warning conditions. The creeks nearest to the Ingalls home were north and northeast of them, thus, it is likely that a south or southwest wind drove the flames on this day.[27]

The Ingalls family employs a common fire protection technique, one that works especially well when conditions are not very windy: they set a backfire between their firebreak and the house. The back-

25. Wilder, *Little House on the Prairie*, p. 277.

26. Wind climatology accessed via the Midwestern Regional Climate Center cli-MATE web tool, available at https://mrcc.purdue.edu/CLIMATE/Hourly/WindRose.jsp.

27. Miller and Rushmore, *Tennessee Wildcat*, pp. 144–45.

fire ensures that the main fire does not have enough fuel to get close to the building. Pa sets the grass outside of the perimeter on fire, with Ma behind him ensuring that the flames do not cross the furrow. The fire surrounds their perimeter with heat and flame and dark smoke; the girls press up against the house while Ma and Pa beat back the fire. They use water and wet sacks to tamp down any flare-ups inside their protection zone. The technique works well enough, and the family is safe, if exhausted.

Pa asks Ma what she would have done if he had been away. "We would have gone to the creek with the birds and the rabbits, of course," Ma replies.[28] Wilder describes how all the animals, small to large, know to go to the creek ahead of the fire. After the fire has passed, the wild things return, first the bigger and faster animals, then the slower moving crawlers.

The Ingalls family's neighbors, Mr. Edwards and Mr. Scott, accuse the Osage Indians of intentionally setting the fires to scare the white settlers who had squatted on their lands, including the Ingalls, Edwards, and Scott homesteads.[29] While it is possible that this was the case, if the Osage did set the fire a more reasonable explanation is that it was a prescribed burn to renew the grasses for new growth. Pa notes that the Osage may have wanted less grass to make it easier for horse travel, while he appreciates that it would now be easier to plow. It is through the lens of fear and lack of understanding about the local climate and ecology, as well as of the Osage neighbors and their practices, that Mr. Edwards and Mr. Scott interpret malicious intent.

In the absence of written records, the true origin of the Kansas prairie fire in *Little House on the Prairie* is unlikely to come to light. The fire could have been controlled or uncontrolled, intentional or accidental, ecological or political. In any case, to the Ingalls family and their neighbors it was unexpected, outside of their control, and frightening.

28. Wilder, *Little House on the Prairie*, pp. 282–83.

29. Ibid.

Minnesota Prairie Fire

While not as ingrained into the ecology of Minnesota as they are in the Flint Hills of Kansas, fires are a natural and routine event on the grasslands of the central Plains, including southwest Minnesota. As with thunderstorms, the weather conditions can be ripe for fires. Though part of the natural rhythm of weather and climate in a region, they can still create damage and hardships.

Plains grassland fires burn swiftly, at lower temperatures than forest fires. They are most common in late summer and early fall, especially in dry years, after the summer sun has baked the moisture out of the tall prairie grasses.[30] While spring fires are common in Kansas, the ground is often wet from snowmelt in the upper Midwest and thus less susceptible to burning.

Before Pa leaves home in *On the Banks of Plum Creek* to work the fall harvest hundreds of miles away, he makes preparations for the family's safety. Fires and the early bite of winter are potential dangers, and he takes measures to prepare for both. He plows a firebreak west of the family's home, over a field where he also sows turnips to provide them a fall harvest of food. There will be no fruit in the aftermath of the locust infestation, but the potatoes have survived, and there are fish in the creek. Pa does not have time to stack hay for animal feed before his departure, but the neighbor, Mr. Nelson, does the chore for him, leaving haystacks near the stable.

On a crisp, clear fall morning—crisp and clear because of low moisture in the air—Laura works extra hard to drive the family's two cattle to meet their daytime herd, because the cows do not want to move toward the west. In the book, Laura smells the smoke, sees the low plume to the west, and recalls the prairie fire in Kansas. In reality, she would likely have been too young during her Kansas stay to have such a vivid connection linking the smell of smoke and a low cloud on the horizon to an approaching grass fire. However, a smell alone can suffice to raise a vivid memory or a trauma response.

West winds carry the smoky scent closer to the house through the morning, until the family can see the glow of the fire beneath

30. Frances W. Kaye, "Little Squatter on the Osage Diminished Reserve: Reading Laura Ingalls Wilder's Kansas Indians," *Great Plains Quarterly* 20, no. 2 (2000): 123–40, http://www.jstor.org/stable/23532731.

the cloud of smoke. By lunchtime, burning tumbleweeds roll ahead of the fire on the winds and across the firebreak, setting fires where they land among the grasses and the haystacks. Mr. Nelson helps Ma beat back the flare-ups with water-soaked rags, while Laura and Mary bring water and wet their rags. Laura, often the brave girl of her stories, also helps tamp down a flare-up next to the house.

The main fire crosses neither Plum Creek nor the firebreak, nor can it back up against the wind or burn over where it has already burned. It withers where it is trapped. Pa's firebreak works.

The experience of the fire near Plum Creek reflects the experience in Kansas almost identically: a plowed firebreak, flames that breach the barrier, and rags or sacks soaked in water and used to tamp down the flames that enter their perimeter. As Wilder would likely have been too young to recall the fire in Kansas, it is no mystery that she uses similar techniques and language to describe the fires in both places. The Ingalls family's experiences during the fire in Kansas likely was recalled to her in family stories, and she may have applied her own experiences from Plum Creek to the family's Kansas experience. This time, though, the Ingalls family relied on help from a neighbor, with Ma admitting they may not have been able to fight back the flames without him.

Homeowners in fire-prone areas today are encouraged to cut back and pull vegetation to leave a fire-safe perimeter around their houses, creating firebreaks just like the homesteaders and their plows. They follow evacuation codes and plans set by local emergency managers and first responders, with systems like "level 1, 2, 3" or "red-yellow-green" activating stages of preparing, readying, and fleeing. The Red Cross provides guidance on ready kits for emergencies like fires, and homeowners plan what they will take and what they will leave behind when they flee. Many in fire-prone areas keep "go bags" packed during the fire season, pre-loaded with emergency necessities such as medications and first aid supplies, shoes, clothes, sanitary supplies, water, food, flashlights, and copies of the family's important documents.[31]

31. National Wildfire Coordinating Group, "PMS 437: Weather: Fire Season Climatology," *Fire Behavior Field Reference Guide* (FBFRG), updated 17

And in the fire season, when they smell that first whiff of smoke from a fire in the distance, they feel the trauma of those past events all over again.

The Prairie Fire and Dust Storm of 1889

South Dakota receives less snow on average than Minnesota. In a particularly dry late winter to early spring, when there is little or no snowmelt or standing water, the ground can be as vulnerable to spring fires as the grasslands of the Kansas Flint Hills, in addition to being vulnerable in a dry summer and in the fall.[32]

If smelling and seeing an approaching fire is viscerally frightening, then smelling it but not being able to see it coming would be terrifying. The distinct and pungent smell of smoke hangs in the air, but a dust storm limits visibility to just a few hundred feet. If a fire is coming—and the smoke foretells that it is—the thick dust enshrouds it like a cloak. By the time someone can see the fire, it will be too late to outrun it, and too late to dig a firebreak that might be ineffective against such gusty winds.

Beginning in April 1889, Wilder's recollection of an early, warm spring in *The First Four Years* is accurate, as the high temperatures in March were well above normal at the first-order station in nearby Huron, South Dakota. The last really cold days were on 13–14 March, though low temperatures still dipped below freezing consistently through March, with a couple of sporadic freezes through early May, ending the freeze season a little early compared to average. March was a dry month, as well, with just 0.19 inches of rain (compared to a normal of about an inch and a half). April and May both had near-normal rainfall, but the rains came mostly on a couple of wet days. From 19 April through 10 May, measurable rain fell on just one day, and that was just a tenth of an inch.

Wilder precisely identified the date of a combined prairie fire

---

Apr. 2024, https://www.nwcg.gov/publications/pms437/weather/fire-season-climatology; Western Fire Chiefs Association, "Wildfire Evacuation Levels Explained," accessed 30 May 2025, https://wfca.com/wildfire-articles/wildfire-evacuation-levels-explained/.

32. National Wildfire Coordinating Group, "PMS 437: Weather: Fire Season Climatology."

and dust storm in the manuscript of *The First Four Years*: 2 April 1889.[33]

Weather records show the high temperature in Huron on 2 April was 70°F, which even today is quite balmy for the location and date. On 3 April, the high rose to just 48°F. A cold front must have passed through, which would have brought a switch to northwest winds. FORTS records support that the winds were very strong on 2 April, and that they were indeed out of the northwest by afternoon. Historical weather maps show that a strong low pressure system moved from northwest Dakota Territory to central Minnesota through the day, with a cold front sweeping through the Dakota Territory.

On the heels of a dry March, the dust from a bare post-winter ground could have been stirred easily by a brisk northwest wind. Dust storms evoke the iconic imagery of the Dust Bowl's black blizzards, with opaque clouds of angry black dirt lurking behind cars, houses, and vulnerable families. Dust storms are not just a product of the Dust Bowl drought, however, and they can happen in dry years on the Plains and in the Southwest, even today. The effects of a black blizzard are just as disorienting as a snow blizzard, the grains even more abrasive than shattered snow crystals. Visibility can drop to near zero, and sandpapery dust flies into the eyes, ears, mouth, and nose.

Vegetation in South Dakota does not usually green up much by early April, even if there are some warm days. The ground cover is the browned, dried-out remnants of last year's plants, and a dry March would have only helped it become even drier. It's no surprise that a prairie fire would be ignited, and with such strong winds, it would be quickly carried quite a distance. With the dust storm limiting visibility and with winds causing such a rapid advance of the fire, the flames would have arrived with little warning to anyone in their path.

In *The First Four Years*, Almanzo and Laura's cousin Peter saves the sheep and douses the stray flames that threaten one of their haystacks, dodging the worst of the prairie fire that burned an extensive

33. CalFire, "Create Your 'Go Bag,'" accessed 17 Apr. 2024, https://readyforwildfire.org/prepare-for-wildfire/emergency-supply-kit/.

path to "the river" (the James or Big Sioux, perhaps?). They re-seed the fields and shear the sheep, and spring moves into summer.

Wilder's fire was not the only one on 2 April 1889.[34] One, ignited by a smoldering straw pile, devastated the new town of Leola and nearly burned into the bigger town of Aberdeen in northeast South Dakota. Another large fire burned across three counties to the west of Kingsbury County (home of De Smet), destroying the town of Ree Heights and just missing Miller, about seventy miles west of De Smet. Another destroyed the town of Mount Vernon, near Mitchell and about sixty miles southwest of De Smet.

The summer of 1889 was the seventeenth driest on record (1881–2024) in Huron, with just 5.21 inches of rain from 1 June through 31 August. If this was supposed to be the year in which Laura and Almanzo decided whether farming was a "success," their fourth year after marriage, they were facing an uphill battle. Because, again, threats and hazards connect together. The drought of the late 1880s created the conditions for a dust storm and prairie fire in April 1889, with cured vegetation and loose, dried-out soil. The crops and trees planted by homesteaders were not made to weather the dry times.

Few threats are as viscerally frightening to humans as that of a fire out of control. The hot, hungry flames and blinding smoke activate our instincts to fight, flee, or freeze. A mere whiff of smoke can invoke a trauma response in fire survivors, as it did when young Laura smells smoke in *On the Banks of Plum Creek*. It can break people down as they watch all their possessions burn, like young married Laura sobbing when her house catches fire in *The First Four Years*.

Fires are fed by a combination of dry ground conditions, hot and dry and windy weather, and vulnerable ecosystems (made so either by natural characteristics or by neglect). Because the antecedent and current weather conditions are enmeshed in fire risk, fires and droughts do often come together. Prolonged dryness primes the ground for fire by drying the vegetation and debris on the ground. Especially prolonged dryness can make trees unhealthy and suscep-

34. National Wildfire Coordinating Group, "PMS 437: Weather: Fire Season Climatology."

tible to burn, while short-term dryness can turn grass to tinder in just weeks. Heat makes the effects of dryness worse.

The connections among heat, drought, and fire have made the western United States especially prone to fires as the climate warms and dries there.[35] Past wildfire management policies and practices of full fire suppression created forest floors with thick undergrowth and dead timber, both creating abundant fuels. Hotter temperatures and lower humidity prime the background conditions for fire to start and spread. The drying of the climate has created stressed, vulnerable stands of trees. When the local winds blow, like the Santa Ana winds in California, fires can spread faster than a car can outrun them.

Drying climates have reached parts of the High Plains as well. The birth regions of the Dust Bowl are prone to cycles of drought and are trending drier, leaving tracts of prairie thick with dead grasses where fires have been suppressed for generations. The swings between wet years and drought years that have characterized the Plains states have set up a pattern of lush vegetation growing in the wet years, then drying into abundant fuels during droughts.

Modern forecasting techniques, dissemination of warnings via phones and sirens, and evacuation plans all help keep humans safe from the fires that encroach on them. Even so, fires can move unpredictably as they create their own weather, respond to rapid changes in weather and fuels, move across peaks and valleys, and are redirected around places where people's homes have only recently encroached. As difficult as it can be to predict winds with precision in complex terrain, predicting the response of a fire to those winds adds another layer of difficulty. As the climate of the future makes areas more susceptible to fire, forecasters cannot predict where those fires will occur or how more science-based wild-

35. Steven M. Ostoja, Allison R. Crimmins, Robert G. Byron, Amy E. East, Michael Méndez, Susan M. O'Neill, David L. Peterson, Jeffrey R. Pierce, Crystal Raymond, Aradhna Tripati, and Ambarish Vaidyanathan, "Focus on Western Wildfires," in *Fifth National Climate Assessment*, eds. Allison R. Crimmins, Christopher W. Avery, David R. Easterling, Kenneth E. Kunkel, Brooke C. Stewart, and Thomas K. Maycock (Washington, D.C.: U.S. Global Change Research Program, 2023), https://doi.org/10.7930/NCA5.2023.F2.

fire management practices will keep people safer. They can predict where the weather conditions may support fire growth more often than they already do now, so that those wildfire managers can take those predictions into account for making plans to keep residents safe.

Weather patterns and climate changes are part of the story of fire threats. Warm and dry weather can create acute fire risk, and the longer-lasting warming and drying patterns from climate change create a chronic fire illness. Trees, grasses, and other vegetation respond to both the longer-term climate and the shorter-term weather in their growth, their drying, and their dying. Humans, too, contribute to fire risks. They build homes in groves of fire-vulnerable trees, light fires in wilderness areas, make mistakes, and make costly choices. They suffer the consequences, too, with the loss of their homes, property, and sometimes lives. The activity of people in fire-prone areas is entangled with fire risk, inseparable, as humans are a part of the environment around them.

10

Grasshopper Weather

On a warm summer day, farmers like Charles Ingalls, laboring over their fields in western Minnesota, straightened their backs to stretch weary muscles. They were grateful for the adequate rains and warm temperatures coaxing the green stems of wheat fields toward the sky—and especially grateful because they had heard rumors that other wheat-growing states had not been so lucky. Drought had been withering the wheat in Kansas and western Nebraska. They had heard faraway news of hungry grasshoppers arriving to devour every green leaf in sight, but so far, the only grasshoppers had been small, just nibbling a few favored garden vegetable plants.

A change in the light caught the eyes of Charles Ingalls and his neighboring farmers. The sky darkened with the shadow of a cloud, like an approaching storm or fire. So often, ominous clouds in the distance foretold of damage and destruction: hail that pulverized crops, fires that devoured them, rains that drowned them. With a million wings bending the sun's light, the cloud cast an eerie shimmer, and the farmers raised puzzled eyes to read and understand the glittering shift in the sunlight.

The grasshoppers dropped from the sky by the thousands, lacking the punch of hailstones but even more destructive to the land. Their bristly, thorny legs raised shivering gooseflesh despite the hot summer sun, imprinting a generation of settlers and their children with a reflexive fear of the sensation of insect legs crawling on skin. Bristly grasshopper bodies crunched under bare feet that had no cleared path, mixing with the slime of grasshopper guts. The chorus of their jaws crushing every plant and plant-like material arose, the volume amplified by the swarm's unimaginable numbers.

The farmers could only stand back and watch, their efforts to kill or clear away the locusts as futile as trying to drive a snowplow through a flood to clear the water. The locusts simply filled in behind, undaunted and dominant in their sheer mass. As Wilder recounted, "I have lived among uncounted millions of [grasshoppers]. I have seen clouds of them darken the noonday sun. I saw their bodies choke the waters of Plum Creek. I saw them destroy every green thing on the face of the earth, so far as a child would know. There are unforgettable pictures of those grasshoppers in my mind that I have tried to draw plainly in *On the Banks of Plum Creek*."[1]

The Ingalls family's fortunes during their first stay in Walnut Grove from 1874 to 1876 were lain at the feet of the "grasshopper" infestation at the center of *On the Banks of Plum Creek*.

The "grasshopper," that is, the now-extinct Rocky Mountain locust, infested wide swaths of the Plains and Midwest in the 1870s. In *On the Banks of Plum Creek*, Wilder invoked all her senses to conjure the infestation to life: the crunch of the grasshoppers beneath their feet, the scratchy feeling of their legs on her body, the chorus of millions of grasshopper jaws crunching through the vegetation, the pungent smell of smoke from smudge fires Pa set in an unsuccessful attempt to keep the locusts out of the wheat.

Wilder was seven years old when the Ingalls family reached Walnut Grove, Minnesota. She likely encountered a few locusts that summer of 1874, but the full invasion arrived at her doorstep in 1875, when she was eight—old enough for the locust swarm to make a permanent and indelible imprint on her senses but young enough for those imprints to traumatize her. Each locust was just

1. LIW to RWL, 17 Aug. 1938, p. 7, Box 13, Rose Wilder Lane Papers, HHPL.

over an inch long and weighed less than an ounce, but the swarm of them put the weight of the world on a young girl's shoulders as they swallowed her family's future with the wheat stalks.

Periodically, the weather and climate conditions contribute to the ideal environment for multi-year locust swarms, something that had occurred before and likely for centuries. So, while the locust infestation would be considered an ecological disaster, rather than purely a weather or climate disaster, weather variations and the background climate of the Plains and Midwest allowed the locusts to thrive and swarm, and the locusts themselves were a native biological component of the ecology of the Plains and Midwest. The difference in the 1870s was the homesteading settlers with farms in the path of those swarms, unfamiliar with the insects' place in the local ecology, whose livelihoods depended on the output of those farms. Disasters occur where humans and hazards intersect.

The Rocky Mountain Locust

The Rocky Mountain locust, *Melanoplus spretus* ("black-armored scorn") is a subset of grasshoppers on the classification family tree, which means that it's not entirely incorrect to call them grasshoppers.[2] Locusts are a kind of uper-grasshopper that bring especially stark devastation by their sheer numbers and mob behavior when they invade. Their ability to swarm and descend en masse distinguishes locusts from more common grasshoppers.

The extinction of the Rocky Mountain locust was unexpected and abrupt, and few specimens were collected of what was perceived to be a widespread scourge before they were gone for good. Most of what scientists know about the locusts is from the observations taken during the mid-1870s swarms and from preserved specimens collected from glaciers in the Rocky Mountains in recent decades, likely deposited there by a wayward transporting wind around the early 1600s and revealed as the glaciers melted in a warming climate.

2. Jeffrey A. Lockwood, *Locust: The Devastating Rise and Mysterious Disappearance of the Insect that Shaped the American Frontier* (New York: Basic Books, 2004), p. 28. Much of the scientific background on the Rocky Mountain locust in this chapter is drawn from this book.

Individually, each full-grown Rocky Mountain locust measured about 1 to 1.25 inches long.[3] Their bodies were brownish-yellow, with markings on the head and body ranging from bluish to purplish to olive green to brown and black—a colorful species. Their wings were distinctly long and spotted. Their smaller nymphs were known to march in a swarm, extending across the prairie in bands as large as a mile across. Rocky Mountain locusts consumed grains, leaves, flowers, berries, stalks, stems, conifers, wood, and even each other—almost anything they could chomp with their jaws.

The Rocky Mountain locust swarms originated from a small "home" space in the river valleys, plateaus, and high plains on the east side of the Rocky Mountains from Colorado to Saskatchewan, especially in Wyoming and Montana. When they swarmed roughly every six to seven years, in irregular cycles peaking in warm and dry years, billions of the insects pushed their range eastward toward the central Plains and Midwest. The locusts then retreated again to their "home" space until the next swarm.

Swarms of the locusts were trigged by both the weather and their chemistry. When the locusts were crowded and denser, they triggered chemical cues that changed the insects' behavior. Far apart, the locusts acted like ordinary grasshoppers; driven together by the search for food during dry weather, they succumbed to a sort of mob mentality, preferring eating over mating, growing longer wings, and launching into swarms. As soon as they depleted the food at their landing site, they launched again. The locust swarms would ride the winds a few thousand feet off the ground and descend again to a new feeding ground.

In *On the Banks of Plum Creek*, Pa wonders aloud how the locusts all know which way was west and their ancestral home, and how they all know to leave all at once. While the locusts are now extinct, scientists have studied similar questions in another migratory insect: the monarch butterfly. The orange-winged butterflies emerge from a winter home space in central Mexico each spring,

3. Charles V. Riley, *The Locust Plague in the United States: Being More Particularly a Treatise on the Rocky Mountain Locust* (Chicago: Rand McNally & Co., 1877), pp. 14–17, https://archive.org/details/locustplagueinunoorile/page/n9/mode/2up.

then migrate thousands of miles into the United States and Canada to feed and breed. Their descendants, hatched from eggs lain on milkweed plants in the spring and summer, migrate back to the warm winter home in the fall. The migratory monarchs have an internal, genetically coded compass wired to interpret the sun's position in the sky and the time of day.[4] The signals combine to tell the monarchs which direction to go and when (southwest, in the fall). Locusts and other migratory insects may also have an internal compass tuned to the sun and time of day to help guide the direction of their expansion outward and return to a home space, as well as chemical or hormonal cues that initiate a swarm as a group.

The locust infestation of the 1870s was far from the first in the region. American Indian and early Euro-American settler records point to infestations in 1818–1820, 1845–1849, 1855–1857, and 1864–1869, with countless others observed by Plains tribes long before the settlers arrived.[5] The infestation in 1864–1869 was the first to threaten homestead settlers, who had faced little weather-related adversity in the few seasons since the Homestead Act of 1862 had opened large tracts of land to farming.

Descriptions of the damage by earlier swarms could have been lifted from the pages of *On the Banks of Plum Creek*. The editor of the *Yankton Weekly Dakotian* wrote on 30 July 1864, "In a short time all were literally covered by myriads of the voracious insects which devoured and destroyed every green thing, even the leaves on the trees and the grass on the prairies. They ate holes in the family washings hanging in the open air and injured many of the tents in which newcomers had made their temporary homes. The insects remained all night and departed the next morning abruptly as they

4. Lockwood, *Locust*, pp. 226–29; Eli Shlizerman, James Phillips-Portillo, Daniel B. Forger, and Steven M. Reppert, "Neural Integration Underlying a Time-Compensated Sun Compass in the Migratory Monarch Butterfly," *Cell Reports* 15, no. 44 (2016): 683–91.

5. Riley, *Locust Plague*; Harold E. Briggs, "Grasshopper Plagues and Early Dakota Agriculture, 1864–1876," *Agricultural History* 8, no. 22 (1934): 51–63, http://www.jstor.org/stable/3739497; and George D. Jones, "Grasshopper Outbreaks in Missouri," University of Missouri College of Agriculture, Agriculture Experiment Station, Bulletin 406 (June 1939), https://core.ac.uk/download/pdf/62792841.pdf; Lockwood, *Locust*, p. 168.

came."[6] A resident of Missouri recalled: "I remember that in the vicinity of St. Louis in addition to their ordinary injuries they stripped the tops of Norway spruce, Balsam Fir, and European Larch; took the blossoms off Lima beans; severed grape stems, and ate numerous holes into apples and peaches, thereby causing them to rot."[7]

And then, quite suddenly, the swarms vanished. The Rocky Mountain locust, the only species of locust in the United States, went extinct sometime around the beginning of the twentieth century; the last known living Rocky Mountain locust was spotted in 1902.[8] While a number of grasshopper species still inhabit and occasionally plague the United States, the locust swarms are relegated to history, and the Ingalls family suffered from their last major swarm in 1874–1876.

"Grasshopper Weather"

Rocky Mountain locust infestations favored unusually dry and warm conditions. Just as cartoon heroes and villains activate their superpowers with a code word, an emotion, or an accessory, the locusts activated theirs chemically, morphing into their beastly selves with extended wings and deep appetites. The locusts gathered, launched, and sought fertile feeding grounds. Once they had stripped the land of anything edible and laid eggs, they again were in close proximity to each other in search of food, whether their feeding ground was hot and dry or wet and lush, and they swarmed again to find food.

While drought conditions could trigger their initial swarm, they were not necessary to maintain it; in fact, they likely would have favored swarming into green areas where there was more to eat, not hot and dry lands with parched vegetation. The ideal weather conditions for the locust swarms may have been drought in their home spaces, but wetter conditions where they landed.

The key to transporting Rocky Mountain locusts during swarm years was the low-level jet, a band of strong winds about 5,000 feet

6. Briggs, "Grasshopper Plagues."

7. Jones, "Grasshopper Outbreaks."

8. Lockwood, *Locust*, explores the ecological mystery of the Rocky Mountain locust disappearance in detail.

off the ground.[9] The low-level jet winds originate in the southern Plains, often in Texas or even in the Gulf of Mexico, and flow north to northeastward across Oklahoma, Kansas, and Nebraska. Though most often observed in the warm seasons, the low-level jet can begin to nudge out of Texas into Oklahoma or its surrounding states almost year-round. When it is present, it is typically strongest in the evening to overnight hours, then loses intensity and shifts its direction during the day, often shifting farther west as it reloads for the next night.

The endpoint, or "nose," of the low-level jet oscillates depending on the weather patterns; it can be as far west as Colorado and Wyoming or as far east as Illinois and Indiana. In an active weather pattern, thunderstorms can develop anywhere along its path. The boost in winds above the ground can cause some of those storms to spin, even providing impetus to produce tornadoes. That said, the nose of the low-level jet can be a weather hotbed; it causes convergence of air at the lower levels of the atmosphere that can often be the focus for shower and thunderstorm development. With origins deep in the south, the low-level jet typically transports moisture deep into the central United States. The moisture it brings, along with the increase in wind shear that these strong winds provide, are key ingredients for the thunderstorms that often develop at the nose of the low-level jet.

In years when the origin region of the low-level jet is dry, it still soars overhead, bringing warmth but little moisture. Without as much moisture to transport, the low-level jet may spark fewer showers and thunderstorms, often only where its nose encounters a richer moisture supply. But there is always something that the low-level jet can move. In addition to water vapor, the low-level jet can transport other small objects or particles, such as smoke, birds, and insects—like locusts. In June 1875, a weather observer in Plattsmouth, Nebraska, about twenty miles south of Omaha, combined his math and weather knowledge to make a critical observation.[10] Dr. Albert Child, an Ohio physician turned Nebraska farmer and

9. Lockwood, *Locust*, p. 23.

10. Ibid., pp. 19–21.

Signal Corps weather observer, noted a swarm of locusts moving northward overhead. Using triangulation, sampling, and comparison sizing, he determined that the swarm was about one-quarter to one-half mile deep, and by telegraphing other observers, that it was about 110 miles wide. Using the mathematics of observing the swarm for five days, moving at around 15 mph, he estimated the length of the swarm at 120 hours multiplied by 15 miles per hour, or 1,800 miles, a number that Child himself noted with an exclamation point. The resulting swarm would have an aerial coverage of 198,000 square miles, which Child deemed "utterly incredible" (a rare adjective in formal scientific observations).[11] Using estimates of locust density compared to modern locust swarms, scientists believe the swarm contained around 3.5 trillion locusts streaming into the Great Plains.

That summer, even though the low-level jet was weaker than normal and even nonexistent for long periods, with its nose snubbed off down in Texas, the locusts were able to surf the wind stream across Kansas and Nebraska and into Minnesota, Iowa, and Missouri. The locusts dropped out of the low-level jet stream and began devouring vegetation and even soft goods like fabrics, leather, wool, and wooden objects. They ate, laid eggs, hatched, and ate more, spreading out of their landing zones as they gobbled the vegetation. Because locusts had dropped out of the low-level jet stream, swaths of untouched land existed in the midst of areas that were fully consumed.

The locusts of the 1870s first reached southwestern Minnesota on 12 June 1874, just fifteen days after the Ingalls family arrived at their land on Plum Creek and before Pa had even filed his claim on the land.[12] According to Caroline Fraser's *Prairie Fires*, the area around Walnut Grove may have missed the first wave of infestation in 1874.[13] It seems reasonable to expect that Pa, an avid consumer of newspaper stories and in-town communication, would have known

11. Ibid., p. 20.

12. R. L. Cartwright, "Grasshopper Plagues: 1873–1877," MNopedia, Minnesota Historical Society, last modified 17 Apr. 2018, https://www.mnopedia.org/event/grasshopper-plagues-1873-1877; Fraser, *Prairie Fires*, p. 68.

13. Fraser, *Prairie Fires*, p. 70.

about the locust infestations surrounding them, and even Laura's character seems to have some awareness of grasshoppers, in her sing-song skipping to the phrase "grasshopper weather" before the locusts arrived.

Rocky Mountain locusts favored spreading during warm and dry weather in their home space, relative to the climate of the late 1800s. In 1874, compared to the 1870s through 1900, unusually warm and dry conditions overspread the central United States. Omaha, Denver, and Cheyenne all recorded their second warmest summer for the 1870s through 1900, while Minneapolis/St. Paul recorded its eighth warmest summer. All these areas would be cooler than normal compared to a more modern climate average, but for the time, it was warm. By the numbers, the summertime average temperature in Cheyenne in 1874 was 67.6°F, compared to an 1873–1900 average of 64.9°F; by comparison, the 1991–2020 normal summertime average temperature is 67.1°F. The summertime average temperature in Minneapolis/St. Paul in 1874 was 70.9°F, compared to an 1873–1900 average of 70.1°F; the 1991–2020 normal summer temperature compares at 71.9°F.

The precipitation conditions were a little spottier in 1874, with pockets of both dry and wet weather. The summer of 1874 was drier than the late-1800s average in Omaha and Cheyenne, while it was wetter than the baseline in Minneapolis/St. Paul and Denver. The dryness around Cheyenne, close to the primary source region of the Rocky Mountain locusts, is important to note. Dry and warm conditions there are consistent with sparking the first wave of the locust infestation into the northern and central Great Plains to the Midwest. Dry conditions in the summer in the central United States are often tied to warmer-than-average temperatures.[14]

In *On the Banks of Plum Creek*, mild winters seemed to be a necessary element of "grasshopper weather," at least in Laura's perception. In reality, the winter of 1874–1875 was not mild at all, as we learned in Chapter 5. In fact, it was a more extreme winter than the noted Hard Winter of 1880–1881, and in particular, it was extremely cold. Wilder probably bent the timeline of events to fit her storyline:

14. 20thCv3 scatterplot, https://psl.noaa.gov/data/atmoswrit/timeseries/.

in this case, that "grasshopper weather" must be mild for grasshoppers to appear the following spring. Locust eggs laid on the Plains were not robust, especially through the winter; most of their numbers originated from their Rocky Mountain genesis region, their surges associated with the warm and dry weather there as well as with the emergence of vegetation elsewhere.

The weather pattern in 1875, in contrast to 1874, was cool and wet across most of the central United States, even in comparison to the relatively cooler averages of the late 1800s. Among six observing stations available for 1874 (Minneapolis/St. Paul, Omaha, Denver, Cheyenne, Bismarck, and Dodge City), only Dodge City registered below-average rainfall compared to the late-1800s baseline, recording its fifth driest year on record. Minneapolis/St. Paul recorded near-average rainfall, and the remaining four sites were all wetter than average. Coming on the heels of the cold winter, 1875 ranks as the coldest year on record in Minneapolis/St. Paul, Omaha, Bismarck, and Cheyenne. It was the fourth coldest in Dodge City compared to the late-1800s baseline and sixth coldest on record, while it was near average compared to the late-1800s baseline in Denver, ranking slightly below average for the whole record. While the cooler and wetter conditions do not fit the typical profile of favored weather for Rocky Mountain locust swarming, they may have already been "locked in" to a swarming pattern for the year, or they may have still encountered pressures in their home space that triggered them to swarm again.

When Albert Childs observed the swarm of locusts over eastern Nebraska in June 1875, the low-level jet winds streamed northward through Texas and Oklahoma to Kansas, then veered eastward into Missouri. On 18–19 June, the low-level jet extended northward through eastern Nebraska and Iowa into Dakota Territory and Minnesota.[15] Otherwise, while Childs counted locusts on 15–25 June, the core of the low-level jet focused south and east of Nebraska. Yet the locusts swarmed across eastern Nebraska through those days, indeed moving "considerably faster than the wind," as

15. 20thCv3 daily, https://psl.noaa.gov/cgi-bin/data/composites/plot20thc.day.v2.pl.

he noted in his observation.[16] Beyond the nose of the low-level jet, winds streamed northward through Iowa into Minnesota.

By 1876, scattered areas of warmer and drier conditions had returned to the central United States. The observation sites closest to the Rocky Mountains were the warmest and driest in the region. Compared to the 1870s-to-1900 baseline, the summer of 1876 was slightly warmer than average in Dodge City and near average in Denver and Cheyenne. It also was on the cool side compared to the late-1800s baseline in Omaha, Bismarck, and Minneapolis/St. Paul. In terms of precipitation, compared to the late-1800s baseline, the summer of 1876 was very dry in Cheyenne and dry in Dodge City, near average in Denver and Minneapolis, and on the wetter side in Omaha and Bismarck. In other words, areas on the eastern and northern side of the Plains tended to run cooler and wetter, while those to the west and southwest tended to run warmer and drier.

The Rocky Mountain locust infestation continued through 1877, with the swarms expanding farther each year than the previous year. The locusts faded away in the summer of 1877, following a smaller hatch.[17] The swarm of the 1870s was the last major swarm of the Rocky Mountain locust. It was too late for the Ingalls family, though. Devastated by the plagues and resulting failed crops, Pa and Ma already were forced to consider their options to make a living, giving up their farm and moving away from Walnut Grove in the late summer of 1876.

Grasshoppers, Poverty, and Relief

What does a family do when it loses everything? In the aftermath of all forms of disasters, the survivors may struggle with physical injuries, trauma and mental health emergencies, and the destruction or loss of their shelter, possessions, crops, and livelihoods.

For many people with tenuous finances, one disaster can be ruinous. Simply rebuilding can be an insurmountable challenge. A damaged shelter must be rebuilt, but the supplies cost money. A

16. Lockwood, *Locust*, p. 19.
17. Cartwright, "Grasshopper Plagues."

lost crop means taking out loans and going into debt just for the seeds to plant the next crop, and if that one fails too, the debt is compounded. Even if the next crop is successful, interest sweeps away part of the profit, and the bills linger forward into the next planting season, and the next. Farmers can find themselves pushed into a deep hole in just one day or just one season, then spend a decade trying to climb out again just to be back where they started before the disaster.

The Ingalls family lost their entire crop, first in 1875 and again in 1876. That they did not lose their crop in 1874 as well, when the locust infestations plagued much of southwestern Minnesota and likely just missed them by a few miles, was a stroke of good fortune.

The locusts raked across large sections of half a dozen states, causing crop failures across entire towns and counties. With the locusts consuming so much vegetation, including crops, gardens, and non-cultivated plants (that is, food growing in the wild and not on farms and homesteads), it was an ecological and agricultural disaster across southwestern Minnesota, southern Dakota Territory (later South Dakota), Iowa, Nebraska, Missouri, and Kansas. Livestock and wild animals alike struggled against starvation without plant feed to sustain them.

While physical harm was rare, the widespread devastation of crops and other property undermined the economic foundation of entire states, especially those populated extensively by homesteaders and small farmers. Neighbors could not turn to neighbors for relief, nor one community to another, as each was as devastated as the next. One of the first widespread agricultural disasters in the fledgling United States, the locust infestation prompted state and federal governments to provide unprecedented relief for their citizens.

The locusts caused an estimated $200 million in crop damages in the United States in 1874 alone, equivalent to about $5.61 billion in 2025 dollars.[18] By comparison, the estimated direct agricultural losses to property during the Dust Bowl years was around $153 million in 1930 dollars, equivalent to about $2.93 billion in 2025 dol-

18. Fraser, *Prairie Fires*, p. 76.

lars.[19] The Dust Bowl contributed a 0.4 percent loss in GDP from 1930 to 1933.[20] The numbers being compared may not be exactly apples to apples, but they highlight the stunning scale of locust-caused losses in the 1870s.

Without the state and federal farm relief that is available today, farmers in the path of the infestation had little means to recover from the devastating multi-year losses of their crops. Farmers who needed work could and did walk dozens or even hundreds of miles to areas that had been untouched to find work as hired hands. This was especially true in the areas where drought had not been raging, itself a threat to crops as much as the locusts who ate them, and that wetter fringe included Minnesota. Whether Charles Ingalls had to walk "200 miles" or a somewhat shorter distance (Laura did tend to overestimate distances), he did join the migrant workers in 1875 and board away from the family home for several months in order to earn enough money to sustain the family through the winter.[21] A distance of about two hundred miles would take him to his brother Peter's farmstead near Zumbro Falls, Minnesota, making the distance plausible, though still an incredible journey by foot. Caroline Ingalls spent his absence pregnant with a son and with worry.

Responding to desperate pleas for food and aid, state officials distributed guidance for eating the grasshoppers, bounties for bushels of dead grasshoppers, and equipment like "hopper dozers" for scooping and killing grasshoppers from fields, measures that all proved no match for either the swarms or the hunger. The proposed measures fell cruelly short of supporting farmers in the path of the locusts, let alone the shopkeepers, laborers, and others who depended on spending by farmers for their livelihoods. States provided seed to farmers via seed grants for replanting, but those were variable among states and inadequate to replace what had been lost.

The devastation was one of the first major events requiring as-

19. Richard Hornbeck, *The Enduring Impact of the American Dust Bowl: Short and Long-Run Adjustments to Environmental Catastrophe* (National Bureau of Economic Research Working Paper 15605, December 2009), https://www.nber.org/system/files/working_papers/w15605/w15605.pdf.

20. Ibid., p. 13.

21. Fraser, *Prairie Fires*, p. 77.

sistance to a large swath of population and sparked decades of debate over how much assistance the government should provide for its citizens in peril. Initially, the states pushed the responsibility of assistance onto the counties, until they recognized that the damage was so widespread that the counties themselves had become impoverished. State governments like Minnesota conflated poverty with moral weakness and laziness and only provided devastated farmers with the bare minimum of assistance to offset outright starvation. While states like Kansas offered direct support, Minnesota first required farmers to liquidate their assets before providing relief, thus ensuring families would not be able to get back on their feet as they sold the necessary horses and equipment. Many farmers, understandably, were unwilling to prove up their poverty for assistance that would be inadequate to meet their family's needs anyway. Further, signing the oath carried a stigma of not being able to support one's family, a walk of shame at the hands of a state government forcing its farmers to publicly humiliate themselves before it would help. Charles Ingalls signed the pauper's oath on 30 November 1875.[22] In exchange for stating that he was "wholly without means," he received two half-barrels of flour.

The losses due to the Rocky Mountain locust infestation had overwhelmed the Ingalls family's tenuous finances and drove them off their Plum Creek farm in search of literally greener pastures. The family filed a preemption claim on their Plum Creek homestead in July 1876, sold it, and staged their departure in the summer of 1876.[23] They first moved east to rest with Peter Ingalls and family near Zumbro Falls, where baby Charles Frederick Ingalls fell ill and died on 27 August 1876. The family buried him there (likely in South Troy, Minnesota) and moved on to Burr Oak, Iowa, in September 1876.

After an ill-fated year in Burr Oak, the Ingalls family returned to Walnut Grove in 1877, then departed in 1879 for good. Wilder omitted these years from the Little House books, sharing no stories of the weather that readers can trace and follow during their year

22. Ibid., p. 77.
23. Ibid., p. 80.

in Burr Oak and the family's second stay in Walnut Grove. When they returned, the family remained in town and rented a home, as Charles and even Caroline and Laura took jobs to earn money. Living in town, the family was not as vulnerable to the whims of the weather. Food sources, schools, and community support were close at hand, if they could find paying work.

Habitats and Humans

Wilder paid attention to the weather and climate, of course, but she also paid close attention to the flora and fauna, the landscape, and the ebb and flow of life around her. Through her observations we hear geese honking high in the sky and splashing into lakes, we smell the violets at the bottom of a buffalo wallow, and we feel the itch of mosquito bites as we slap them away in the still summer air. She might have needed help from her sister Carrie and aunt Martha to label the flowers and birds, but she recalled the nuances of their color and size, their seasonal arrival and departure, their smell and taste.

She and Pa noticed that as people built towns and homes, the animals departed. The large animals left first: wolves, deer, and antelope found safer homes farther away from the increasingly dense human settlements. In *By the Shores of Silver Lake*, when Pa comes home empty-handed from a hunt in an area that had become more settled, he notes that it would be harder to hunt geese.[24] Many modern suburban dwellers will note that geese make their homes among residential neighborhoods and business districts quite readily. Perhaps the geese have adapted to the noise of humans, or perhaps the food discarded by humans is enough motivation to overcome their apprehension.

Ironically, human activity almost certainly caused the extinction of the Rocky Mountain locusts. The locusts, as we have seen, originated from the flat, elevated plateaus of the Rocky Mountains, mainly in Wyoming and Montana. Such a small and specific region of origin can make any species vulnerable to disruptions in that region. It is no coincidence that the last Rocky Mountain locusts were

24. Wilder, *By the Shores of Silver Lake*, p. 245.

documented in 1902, after Western agricultural and ranching practices moved into the Rocky Mountain plateaus. Settlers churned up the locusts' eggs as they plowed the ground and changed the water patterns of the river valleys by diverting streams for irrigation. As much fortitude as the swarms had enjoyed on the Plains, they were rather vulnerable in their home space. When their ecosystem, balanced on its narrow foundation, teetered and then toppled, the species collapsed.

Because the Rocky Mountain locusts were not well studied before their demise, scientists remain unsure of the ecosystem impacts of their extinction.[25] Was their extinction a "good" or "bad" thing? There is not enough scientific evidence to answer objectively, though the answer from those who invaded their spaces and no longer need to worry about a catastrophic swarm is clearly on the side of "good." We simply don't know what other parts of the native prairie ecosystem relied on the locusts for food, for clearing undesired or unwanted vegetation to make way for new growth, or for fertilization, nor do we know what other species of predator or invader thrived because of the locusts' extinction.

Just as human land use can disrupt ecosystems, so can human-caused changes to air and water. The fingerprint of human activity is pressed into each extra molecule of carbon dioxide in the atmosphere released by burning oil, gas, and coal to heat and power our homes and businesses. Each molecule is an extra source of energy in the atmosphere, basking in the sun's radiance and emitting it enthusiastically in all directions: toward other molecules, toward the land, and out to space. All that extra energy influences weather patterns, evaporates water into the air (which becomes even more energetic than carbon dioxide molecules), melts ice and snow, and nudges the entire atmosphere and ocean system to be warmer. Every day, every winter, every year is not smoothly warmer than the last; the trend bumps up and down like the value of a dollar through the centuries. But its direction is clear and distinct, ticking ever upward.

As the weather patterns change, so do rainfall patterns, perhaps

25. Lockwood, *Locust,* p. 137 and Ch. 14.

increasing or decreasing, perhaps shifting from one season to another. Temperature patterns change, and the kinds of cold that used to sweep through and freeze out some of the bugs and pests become less common. Now those bugs and pests can lay eggs and linger, shift north, spread their diseases and crop damage even farther. Trees that stood for centuries (where they had not been cut down by encroaching humans) shrivel in the hotter and drier summers, inviting invaders into the chinks in their armor, then stand desiccated and helpless as wildfires burn through them.

The ties that bind weather and climate to the plants and animal life are not just knotted and braided together, they are welded fast. All manner of critters follow the conditions that support them, including the food they need to eat, clean water to drink, and temperatures that let them maintain their body temperatures and raise their babies. The activities that sustain modern human life push the weather and climate and pressure the landscape. Sometimes this pressure eliminates a pest like the Rocky Mountain locust. Sometimes it threatens the very creatures that help modern humans survive and thrive, like monarch butterflies and other pollinators. We are inextricable from the ecosystem around us, even when surrounded by the thick walls of modern homes and nestled into neighborhood enclaves.

The Rocky Mountain locust catastrophe had roots in many contributing factors. Weather patterns set up dry and hot weather in the origin region of the locusts and adjacent swaths of the Great Plains. The pressures on the locust population put in place by the weather triggered a biological response in them that shifted them into "mob mode," transforming them into swarming machines. Human encroachment in their swarm zones put people's crops and livelihoods into the path of harm, with the new residents relying heavily on the health of crops that were distinctly vulnerable to the invasions. The weather patterns at the origin of the swarms lingered for years, and the swarms spread as the locusts descended, ate, laid eggs, and swarmed again. The resulting disaster was widespread and long-lived, drawn out over several years and numerous states and territories.

People took many blows from the locust infestations. Foremost and at the root of many other impacts, the locusts hit them in the pocketbooks, devouring their livelihoods with each stalk of wheat and garden sprout. Without adequate food, many residents teetered on the brink of starvation and certainly suffered from nutrition deficits, impacting their overall health and wellness. The locusts invaded the mental health of the residents, too, creating untold harm to family safety and stability. Like the Ingalls family, many residents lost their homes and land. As a direct result of being upended by the locust infestation, the Ingalls family found itself in its most unsafe situation of Wilder's life: the year in Burr Oak, Iowa, from 1876 to 1877, spent in close quarters with drunken brawls, spouse abusers, liquor-soaked fires, and threats to Wilder and likely other children.[26] The Ingalls family would be forced to flee Burr Oak in the night to escape debt.[27]

Humans did not bring about the extinction of the locusts any more consciously than the locusts intentionally attacked humans. In both cases, the populations of one group were simply in the way of the other. The difference is that the locusts had cycled through their swarming grounds for a very long time and coexisted with humans before agricultural settlement, while humans invaded the homes of the locusts and settled where they had not lingered before. In this case, it was the humans who were the invasive species, and human encroachment in previously balanced areas now threaten to push the monarch butterflies to a similar fate as they are thrown out of ecological balance.

People are not separate from the ecosystems around them. They are a part of the ecosystems and still depend on their health for their own well-being. Humans, like other animals, need clean air and water, healthy vegetation, and balanced animal populations.[28]

The imprint of humans on the world around us is unmistakable and irrefutable. Humans have changed the vegetation that grows, the paths of rivers and shapes of lakes, the living creatures that thrive

26. Wilder, *Pioneer Girl: The Annotated Autobiography*, pp. 99–106.

27. Ibid., p. 112.

28. Robin Wall Kimmerer, *Braiding Sweetgrass* (Minneapolis: Milkweed Editions, 2015), p. 166.

or die, and even the chemical composition of the air we breathe and the water we drink. Sometimes the changes have very small effects, while others change the landscape forever. Some changes result in a net benefit to the humans around them, while others are detrimental, and a few are neutral. The moral debates around how much humans should shape the world around them for their own comfort, convenience, and livelihood or leave the environment intact are emotionally charged. We struggle with questions about who should be in charge of the decisions, who should pay when human intervention costs the people who live in the affected area, how to restore ecosystems to their pre-industrial states, and how much human burden on the environment is too much.

Answering questions like these starts with paying attention to the wider world—more than just the weather conditions, even. The weather and climate connect to our lives not just through our safety from weather hazards like fires, floods, and storms—they also connect to our financial stability. Scientists can measure the changes, collect observations from scientists and citizens alike, analyze them, and report them. With the collection of these records and stories, others—policymakers, residents, leaders—can begin to make their cases for what to do with that information. They can make decisions about community planning to build in resilience to disasters, relief for those affected, and mitigation to prevent future catastrophes.

11

An Ideal Climate

A young couple drew their wagon to a halt in front of a luscious green hill near Mansfield, Missouri.[1] After a six-week wagon journey and many more days of searching, they laid their weary eyes on the land their hearts desired. A cool, clear spring ribboned through the land, descending from a hilltop. Unplanted apple trees lay in wait to sink their roots into the fertile soil. A log cabin welcomed their family, in need of cleaning, but cozy and inviting, at least for the time being. A space of land at the edge of the hill, shaded by an old oak tree and sloping down to the ravine on one side and up toward the rocky, heavily wooded hill on the other, beckoned the couple to build their own home. Their farmhouse would be sized and spaced for their needs and constructed from materials found on their own land. The spring running nearby would provide water until they could dig a well and install a pump.

The Wilders had purchased this parcel of land and

1. Details of this section are drawn from Laura Ingalls Wilder, "On the Way Home" in *A Little House Traveler: Writings from Laura Ingalls Wilder's Journeys Across America* (New York: Harper Collins, 2006), pp. 88, 98–103. The weather reports are from official weather observations at Springfield, Missouri, and Huron, South Dakota. The insights into what the Wilders may have been seeing or feeling are informed but fictionalized.

soon named it Rocky Ridge Farm. It would be home for the rest of their lives.

Temperatures had been warm in the few days after the Wilders arrived in Mansfield on 29 August 1894, but the shade from the abundance of trees and the gentle breezes had made the temperatures in the eighties feel tolerable. After so many dry years in South Dakota, the rains that fell on 2–4 September, at least as measured in Springfield, must have washed away the memories of Dakota drought and dust. A downpour of over three inches of rain on 10 September, on the heels of two days with temperatures in the nineties, ushered in a drop to much cooler temperatures in the sixties for several days in a row. The mild and wet weather laid a balm on their bodies as temperatures stayed in the upper seventies to lower eighties for the rest of the month.

The winter ahead was on the cold and snowy side, relative to the averages for the time in southwest Missouri. But it was far milder than a bracing Dakota winter, even though the winter near De Smet was close to its average for those years. Temperatures routinely stretched above freezing in Mansfield. Even though nearby Springfield actually saw more snow than De Smet during the winter of 1894–1895, it didn't last as long on the ground. And snow meant moisture, which meant water for the roots of all their abundant vegetation in the spring. The lack of snow near De Smet only meant that it was still dry, that farming there would still be tortured.

Mr. Wilder found the relief he sought for his aching body. Mrs. Wilder found the pleasantly mild air interspersed with the occasional excitement of storms to be just enough but not too much variety. Both saw the abundance of green plants and fresh water, beckoning them to lay down their own roots as they planted the roots of seedling apple trees and crops to come.

Mr. and Mrs. Wilder had uprooted their lives and moved away from everything and everyone they loved to find a home that soothed their aching bones and hearts. They were hardly the first or last family to migrate for health.

Weather, Climate, and Health

The field of research connecting human health to various weather and climate conditions is as extensive as it is still rife with gaps in our understanding. Researchers well understand that heat waves are deadly, especially to vulnerable people such as the elderly, infants, and those with chronic illnesses. They have identified the spread of tropical diseases as the habitats of their vectors spread in a changing climate. People swear they can feel a storm coming in their joints or headache patterns, though this connection is more tenuous than others, at least in the research.

In *By the Shores of Silver Lake*, Wilder introduces the concept of moving to a better climate for health. As the Ingalls family settles in for the winter, Pa encounters a stranger who tells of a man named Woodworth, sick with consumption (now known as tuberculosis). Woodworth "came out here to take the prairie-climate cure" but was too sick to continue to live in his cabin through the winter.[2] Pa and the stranger load Woodworth into a wagon, bound for the home of relatives in the East for the winter. The same Woodworth becomes the depot manager in De Smet in later books, and his son, Ben, hosts Laura's first teenager party in *Little Town on the Prairie.*

In the late 1800s, recommendations for fresh air and rest as a cure for consumption were not uncommon, with patients often seeking a different climate, respite at a sanatorium, or both.[3] Even in the late 1800s, research had uncovered the bacterial cause of consumption, but the cure still evaded doctors at the time. The "prairie cure" was not entirely unfounded; being in open air did slow the spread of tuberculosis, as it does many airborne illnesses, as well as provide some symptom relief. Antibiotics have greatly reduced exposure to and risk of tuberculosis in modern America, regardless of climate.

Following a nearly fatal bout with diphtheria in the spring of 1888, along with a complicating stroke during his prolonged recovery period, Mr. Wilder lived the rest of his life with stiffness and,

2. Laura Ingalls Wilder, *By the Shores of Silver Lake* (1939; New York: Harper Trophy, 1971), p. 151.

3. Sheila M. Rothman, *Living in the Shadow of Death: Tuberculosis and the Social Experience of Illness in American History* (New York: BasicBooks, 1994).

presumably, pain. As a stoic farmer, he did not verbalize his pain or symptoms, and Mrs. Wilder did not air them, save for small hints in her writing. In her article "Our Little Place in the Ozarks," Wilder wrote, "We were looking for a place where the family health might make a good average, for one of us was not able to stand the severe cold of the North, while another could not live in the low altitude and humid heat of the Southern state."[4]

Many people prefer a certain climate, something that feels the most comfortable, or the most familiar. People who grew up in the northern United States may feel off balance if their year does not include a true winter season, with snow and cold temperatures. Others from the same areas may wish to flee the cold and snow of their younger years, seeking milder climates to thaw their bones. Those who grew up in drier Western climates may feel drowned by the high humidity of the eastern United States; those who grew up in the East may feel shriveled by the dryness of the West. The loud, violent storms of the Plains may alarm people who did not grow up experiencing them, and they may frighten some people who did grow up with them into moving out of their range. Just about every corner of the United States has some version of hazardous weather that threatens the people who live there.

Even in the modern world, people's relationship between their health and the weather and climate around them can influence decisions on where they live. People move to avoid allergy triggers, seeking to alleviate asthma, hives, or ear, nose, and throat irritation. Some move away from hurricane-prone coasts and tornado-prone cities because the trauma affects their mental health. Drought is linked to increases in mental health concerns and deaths by suicide in farmers.[5] Those with seasonal affective disorder may move away

4. Laura Ingalls Wilder, "Our Little Place in the Ozarks," *Missouri Ruralist*, Dec. 1923, in *Laura Ingalls Wilder, Farm Journalist: Writings from the Ozarks*, Stephen W. Hines, ed. (Columbia: University of Missouri Press, 2008), p. 24.

5. Jesse D. Berman, Marizen R. Ramirez, Jesse E. Bell, Rocky Bilotta, Fredric Gerr, and Nathan B. Fethke, "The Association Between Drought Conditions and Increased Occupational Psychosocial Stress among U.S. Farmers: An Occupational Cohort Study," *Science of the Total Environment* 798 (Dec. 2021), https://doi.org/10.1016/j.scitotenv.2021.149245.

from cloudier climates like the Great Lakes and the Pacific Northwest to improve their ability to function.

Growing up in Michigan, cloudy winters were my normal. Only when I moved out of the state, to the Washington, D.C., area, then to Iowa, Nebraska, and Oklahoma, did my winters become progressively less cloudy. Only then did I realize how important it is for me to see a little sunshine between the storms during the winter, to have a thaw from time to time. My choices to move were not driven by health, but the moves did affect my health. Sunny and warmer winter days have moved me out of hibernation and into a healthier level of activity for my body. Though I do not suffer from seasonal affective disorder, seeing the sunshine and blue skies does lift my mood.

Not everybody has the means to move to a new location where the weather and the climate are better suited to their physical and mental health. Additionally, the process often involves trading one type of stressor for another. As the saying goes, the grass isn't always greener on the other side. In the modern world, we have resources to help us adapt where we are, from heating and cooling systems to phototherapy and allergy medications. While these measures don't remove all discomforts, they can make it easier to live with them.

Mr. and Mrs. Wilder specifically included their health and comfort in their decision to move to Missouri. With Mr. Wilder's lingering health concerns, a suitable climate was especially important for his ability to function as a farmer. Mrs. Wilder's comfort was a factor, too, as their brief foray in Westville, Florida, made clear that a fully tropical climate was not suitable for her. Their health and comfort were not the only factors; flyers about Missouri drew them with their promises of green lands and fertile fields—that is, adequate precipitation and less tendency for drought. Greater stability and less variability in the whims of weather were important to the Wilders, too, who had been burned by drought, heat, and hail in their attempts to farm in South Dakota. Few people make the decision of where to live based solely on health and comfort; jobs, safety, lifestyle, recreation, and even scenery are just a few of the other considerations.

The Wilders' Ideal Climate

Laura Ingalls Wilder wrote of her and Almanzo's selection of a forever home in Mansfield:

> It was rather unpleasant journeying in the heat of the summer, but as we climbed into the hills this side of Springfield, the air grew fresher and more invigorating the farther we went until … we found the place we were seeking, far enough south so that the winters were mild; high enough for the air to be pure and bracing; sheltered in the hills from the strong winds of the west, yet with little breezes always blowing among them; with plenty of wood for fuel and timber and rocks for building; with low lands for cultivation and upland bluegrass pastures for grazing; with game in the woods and fish in the rivers; and springs of pure, cold, mountain water everywhere.
>
> Here on the very peak of the Ozark watershed are to be found good health, good homes, a good living, good times and good neighbors. What more could anyone want?[6]

While the climate of southwestern Missouri is more humid than that of the Dakota prairies or the Minnesota valleys, it also faces fewer temperature extremes, especially on the cold side of the temperature spectrum. In fact, the humidity is part of the reason why the temperature is less variable. Water gains and loses heat more slowly than air, so the more humid the air, the smaller the temperature swings. A foggy day and night provide an extreme example; temperatures may change only a few degrees from high to low, while temperatures on a clear day and night at the same location may swing twenty to thirty degrees or more. A city in a tropical climate is likely to have higher average nighttime temperatures and lower average daytime temperatures than a city in an arid climate at the same latitude, and the record highest and lowest temperatures will both be more extreme at an arid location than at its tropical counterpart.

In today's climate, it is still clear that temperatures in southwest-

6. Laura Ingalls Wilder, "Our Little Place in the Ozarks," *Missouri Ruralist*, Dec. 1923, in Hines, *Farm Journalist*, p. 24.

ern Missouri are simply less extreme than in eastern South Dakota. The difference between the highest (July) and lowest (January) normal high temperatures in Springfield, Missouri, is 46°F, while the difference in Huron, South Dakota, measures at 61°. The difference between the record warmest high temperature (113°) and record coldest low temperature (-29°) in Springfield is 142°, compared to a whopping 155° in Huron (record warmest temperature of 112° and record coldest temperature of -43°).

The warmest temperatures in Springfield and Huron are similar, especially the extremes. The difference between the two locations is more apparent in the normal and extreme cold temperatures, both of which are quite a bit warmer in Springfield than in Huron. Winter temperatures in southwestern Missouri are far milder, on average, than in eastern South Dakota, while summer temperatures are only somewhat warmer. The result is less swing from the highest temperatures to the lowest, on average, through a year.

Precipitation is also less variable in southwestern Missouri than in eastern South Dakota. Farmers can better plan for their plantings when the precipitation amount is more stable and reliable. At Springfield, the record wettest year was 150 percent wetter than average, and the record driest year was 40 percent drier than average. In Huron, the wettest year was 184 percent of average, and the driest year was just 26 percent of average. In other words, the highs are higher and the lows are lower in the eastern South Dakota climate.[7]

Plants and animals alike are more comfortable under less extreme weather. Hot weather can cause people and animals heat stress, heat exhaustion, and heat stroke, especially if it is also humid. Extreme cold weather can cause frostbite and hypothermia, especially if it is also windy. People do acclimate to the climate where they live, but extreme heat and cold are dangerous nonetheless. Crops suffer with

7. "Average" in this case is the average for the full period of record at Springfield (1888–2024) and Huron (1881–2024), calculated from raw observed data obtained from the ACIS database. This is in contrast to the "normals," which are a baseline comparison period of 1991–2020, and also in contrast to homogenized climate data, which are quality-controlled for station moves and other inconsistencies. The choice to use a raw average here is for simplicity of comparison; anybody can access the raw observed data and make a comparison.

too much or too little water, untimely heat or freezes, or even lack of freezes, depending on the plant. Each living being has a narrow range of what is comfortable and a wider range of what is tolerable.

As climate changes, it nudges the conditions away from those ranges, creating uncomfortable and even intolerable climates for the plants and animals (including humans) that used to thrive in that very same location.

Home Feels Different Now

Wilder's readers and fans who wish to replicate her family's life experiences can visit each homesite along the Ingalls and Wilder families' paths: Pepin, Malone, Independence, Walnut Grove, Burr Oak, De Smet, Spring Valley, Westville, and Mansfield. They can follow the trail of breadcrumbs, viewing the homes or replicas, the historical markers, the plates that Wilder used, the quilts she pieced together. They can cook the foods she cooked and described, make and wear garments like hers, see pictures of her and her family. They might see a tree that Pa planted or a well that he dug. They can even hear Pa's fiddle.

Wilder's readers and fans can and do experience weather elements that Wilder and her family experienced, hopefully from places of relative safety and comfort as modern sheltering and warning systems allow. They may see the whirling whiteness of a blizzard and feel the sharp sting of subzero-degree air inhaled deep into their noses and lungs. They can see tornadoes, hear thunder crashing at night, and pick up icy hail after a storm. They can smell a prairie fire or see the flickering red flames of distant forest fires. They can feel the grit of dust driven on the wind during a drought. They can hear the rushing water of a flood and see the paths and roads covered by swollen rivers.

But Wilder's fans will never experience the climate of each site as Wilder experienced it. Those climates are gone, lost in the sweet by-and-by. The individual weather events may still happen where they happened in her stories, but their frequency and timing have changed. Climate is the average, the sum, of weather over a long period of time. Those averages and the sum of all weather have changed enough at every one of the Ingalls and Wilder homesites that it is measurable by instruments. But more than that, it is noticeable by the people who live there.

The change in weather is observed by farmers and ranchers in the crops they can sustain and still make a profit on, the amount of water they use and the depth of their wells, the number of years with lost crops or livestock and the reasons why they were lost. Farmers and ranchers deal with new pests and invaders as the collection of weather shifts. Insects and small animals, predators and prey, even some weeds and bushes and trees migrate farther north, change their patterns, come earlier and leave later.

Heat has crept in as a weather hazard at all the Ingalls and Wilder homesites, and it is now the deadliest weather hazard in the United States. It is especially hazardous where people are less acclimated to it—places in the northern half of the country where heat is still more rare, for now, and more of a shock to human bodies. In some parts of the United States, such as Wisconsin and upstate New York, many homes and buildings lack air conditioning, even while heat waves are increasing in both frequency and magnitude. Heat certainly has increased in southern Missouri, where both temperatures and humidity are climbing.

Wilder rarely mentioned being too hot in any of her books or manuscripts.[8] Other than her time in southern Kansas, she spent her childhood in northern climates, north of 43 degrees latitude (the latitude of Milwaukee, Boise, and Syracuse), where winters bring cold and snows, summers are warm but not hot, and fall and spring are battlegrounds between warm and cold. She grew up in and was acclimated to a continental climate, where the changes of seasons are distinct.

In fact, one can wonder if the deep chills of Wilder's childhood inclined her to desire warmth in her adulthood. The Florida climate, as we have seen, was too warm and humid for her, but the climate of Mansfield was quite a bit warmer than those of her childhood homes. Perhaps it took a lifetime to thaw after the Hard Win-

8. John F. Case, "Let's Visit Mrs. Wilder," in *A Little House Sampler*, ed. William T. Anderson (Lincoln: University of Nebraska Press, 1988), p. 9.; and Fraser, *Prairie Fires*, p. 163.

ter of 1880–1881, the deep cold of 1874–1875, and the many winters in Wisconsin, Minnesota, and South Dakota.

In a generation or two, if climate keeps changing at its current pace, the weather in places like Pepin and Walnut Grove will resemble aspects of the mild weather of Mansfield in Wilder's days there more closely than it will resemble the weather of Wilder's own youth at those sites. A traveler would need to go to the northern parts of those states to feel the same wintry weather that Wilder did as a girl.[9] In just a couple decades, if climate keeps changing at its current pace, the winters that used to be in places like Kansas City, Omaha, Des Moines, and Moline will shift northward into Wilder's childhood homes.[10] In another fifty to a hundred years, the summer temperatures in Wilder's childhood homes may resemble those that used to be in places like Missouri, Kansas, Arkansas, and Louisiana, and a traveler might need to be in Canada to experience the same types of winter weather that Wilder experienced in her childhood. Some of the most extreme warming in the lower forty-eight states will be in the north-central states like Wisconsin, Minnesota, and the Dakotas. Rather than people migrating toward warmer weather, the warmer weather is migrating to them.

To be sure, it still gets cold at Wilder's childhood homes. Winter is still winter. Temperatures still dip below zero, snow still falls, and blizzards still rage. The grip of winter, though, is just a little more tenuous. The thaws between the freezes last just a little longer. The most extreme cold comes less often. The hardest winters come once in a generation instead of once or twice in a decade. Spring knocks

9. "IPCC, 2021: Summary for Policymakers" in *Climate Change 2021: The Physical Science Basis. Contribution of Working Group I to the Sixth Assessment Report of the Intergovernmental Panel on Climate Change*, eds. V. Masson-Delmotte, P. Zhai, A. Pirani, S. L. Connors, C. Péan, S. Berger, N. Caud, Y. Chen, L. Goldfarb, M. I. Gomis, M. Huang, K. Leitzell, E. Lonnoy, J. B. R. Matthews, T. K. Maycock, T. Waterfield, O. Yelekçi, R. Yu, and B. Zhou (Cambridge and New York: Cambridge University Press, 2021), accessed 8 May 2024, https://www.ipcc.ch/report/ar6/wg1/.

10. Kaitlyn Trudeau, "Shifting U.S. Cities," *Climate Central*, 13 July 2022, https://www.climatecentral.org/climate-matters/shifting-u-s-cities; University of Maryland Center for Environmental Science, "What Will Climate Feel Like in 60 Years?," https://fitzlab.shinyapps.io/cityapp/.

on the door just a little earlier. Trees emerge from hibernation earlier, green-up arrives earlier, and ice breaks up earlier. It is not a smooth change through time. After a couple of mild winters, a severe one sneaks in. Winters that end early mix in with those that end later. A little like the value of someone's home or retirement savings, the year-to-year numbers can fluctuate, but the trend is distinctly and measurably upward over time.[11]

Wilder's readers reminisce about decades and centuries past. Little House fans view the changes in clothing, food, furniture, and transportation with wonder. Wilder herself witnessed the changes from horse-drawn buggies to trains to automobiles and airplanes, from quill-and-ink to pens to typewriters. The collection of weather in the past, compared to today, is another source of nostalgia. In awe, we can reflect on the advance of science that allows meteorologists to make forecasts, issue warnings, and observe the weather, from backyard gauges to satellites and radar and powerful computers. With longing, we look back on the weather of Wilder's years and the current collection of weather, and with curiosity and often concern, we ponder what the collection of weather in the future will bring.

"The Great Woods Have Been Destroyed"

From the character of the land cover to the ice-out dates at Lake Pepin, the fingerprints of human activity on local and regional climate are distinct and observable. In her 1919 article "The Great Woods Have Been Destroyed," Wilder focused on the future of energy and natural resources in the United States, specifically deforestation, a broad and weighty topic (in that era) for a Missouri farm woman's thoughts.[12] She had seen deforestation with her own

11. K. Marvel, W. Su, R. Delgado, S. Aarons, A. Chatterjee, M. E. Garcia, Z. Hausfather, K. Hayhoe, D. A. Hence, E. B. Jewett, A. Robel, D. Singh, A. Tripati, and R. S. Vose, "Ch. 2. Climate Trends," in *Fifth National Climate Assessment*, eds. A. R. Crimmins, C. W. Avery, D. R. Easterling, K. E. Kunkel, B. C. Stewart, and T. K. Maycock (Washington, D.C.: U.S. Global Change Research Program, 2023), https://doi.org/10.7930/NCA5.2023.CH2.

12. Laura Ingalls Wilder, "The Great Woods Have Been Destroyed," *Missouri Ruralist*, Sept. 1919, in Hines, *Farm Journalist*, pp. 76–77.

eyes in her childhood, as she wrote: "Too many of the forests of the United States have been made into lumber even though there never has seemed to be lumber enough, and the waste of timber has been great. The great woods of the East and the North of our country have been destroyed."[13]

Presciently, Wilder stated that "the end of hard coal is in sight" and "thus might be created rivers of power flowing through the country and furnishing energy and power to our manufactories [*sic*] at much less than half of what it costs now."[14] Advocating for renewable energy sources and a nationwide power grid, she feared a day in which the country would run out of fuels to generate electricity, noting that "it need never arrive if we see to it that our water power and what is left of our fuel supply is handled carefully and intelligently."[15]

As Wilder had pointed out, the native forest across large swaths of the northern and eastern United States was cleared for agriculture and towns in the nineteenth and early twentieth centuries by Euro-American settlers. American Indians had also cleared forested lands to create fields for crops.[16] Meanwhile, during the Homestead Act era, settlers were encouraged to plant trees in "tree claim" parcels in the Plains, where they were simply not sustainable.[17] The federal government advocated for planting trees because of the influence they would have on the climate of the area, providing shade and windbreaks and, in their prognostications, altering precipitation patterns. Following the Dust Bowl of the 1930s, the federal government again advocated for planting trees where they did not previously exist, this time in the form of shelterbelts that would break the wind and allow dirt to settle, a practice that was far more scientifically supported than the assumptions underlying tree claims.[18]

13. Ibid., p. 76.

14. Ibid., p. 76.

15. Ibid., p. 77.

16. *See* William Cronin, *Changes in the Land: Indians, Colonists, and the Ecology of New England*, rev. ed. (New York: Hill and Wang, 2003).

17. C. Barron McIntosh, "Use and Abuse of the Timber Culture Act," *Annals of the Association of American Geographers* 65, no. 3 (1975): 347–62, http://www.jstor.org/stable/2561886.

18. Egan, *The Worst Hard Time.*

Changing the land cover does indeed affect the local climate.[19] Urbanization causes distinct changes in weather, with urban heat islands and increased precipitation downwind of major urban centers, but even changing the type of vegetation influences local climate. Converting forest fringes to agricultural land causes cooling temperatures, especially daily maximum temperatures, in parts of the Midwest in the winter, summer, and fall, while temperatures increase in the spring. Deforestation also reduces evapotranspiration, the release of water vapor into the air by plants, in the fall and spring; evapotranspiration increases in the summer, when crop growth peaks. Dew point temperatures, or moisture in the air, due to evapotranspiration also tend to increase over areas where native grassland is converted to agriculture.

Human-caused climate change is dominated by the effects of increasing carbon dioxide in the atmosphere, a byproduct of the burning of coal and wood, as Wilder alluded to in her article, as well as oil-based fuels, natural gas, and other fossil fuels. It is so pervasive that it underlies all weather in the atmosphere. Whether we can detect the influence or not depends on other patterns in the atmosphere at the time, the strength of the climate change signal, and the limits of what we can measure in the atmosphere. The ice cover records in Pepin provide a clear example of the effects of climate change, as the average ice-out date has migrated from 12 April to 29 March, a two-week shift, since recordkeeping began in 1849.[20]

While temperatures are warming year-round, winter temperatures have borne the brunt of change. This is a hallmark of climate change: the coldest times and places warm faster than warmer times and places. This means that in most places, the Arctic is warming faster than the tropics, winters are warming faster than summers,

19. P. E. Thornton, B. C. Reed, G. Z. Xian, L. Chini, A. E. East, J. L. Field, C. M. Hoover, B. Poulter, S. C. Reed, G. Wang, and Z. Zhu, "Ch. 6. Land Cover and Land-Use Change," in *Fifth National Climate Assessment*, eds. A. R. Crimmins, C. W. Avery, D. R. Easterling, K. E. Kunkel, B. C. Stewart, and T. K. Maycock (Washington, D.C.: U.S. Global Change Research Program), https://doi.org/10.7930/NCA5.2023.CH6.

20. Pepin ice-out data from "Lake ice in/out report: Pepin (25000100)," Minnesota Department of Natural Resources, accessed 19 Feb. 2024, https://www.dnr.state.mn.us/lakefind/ice.html?id=25000100.

and nights are warming faster than days. At La Crosse, Wisconsin, average temperatures have warmed by 2°F for December through February through its period of record, 1872–2025. When we think about day-to-day weather, 2° sounds small. But seasonal temperatures don't feel the same and don't respond the same as daily temperatures. It takes big changes to move the average of an entire season.[21]

More importantly, in most cases, it isn't actually the change in average temperature that causes impacts. The big worry is that when the average changes, it also affects the chances at the extremes—making extreme cold events less likely and extreme warm events more likely. For many residents of the upper Midwest, a little extra warming in the winter sounds more pleasant than problematic. The warm spells, though, can allow trees and plants to begin to bloom early, only to be nipped later by seasonably (in other words, not unusually) cold temperatures. The lack of extreme cold can allow pests, diseases, and fungi from warmer climates to expand their reach northward, impacting ecosystems that are not adapted to their presence.

Precipitation signals are less coherent, but they still emerge in many locations, including a trend toward wetter conditions in the upper Midwest to the Great Lakes and drier conditions in the southwestern United States. In particular, winter and spring are trending wetter. Temperatures are warming in the winter, but they are still cold enough for snow. Thus, perhaps counterintuitively, average snowfall is increasing in some places despite the warming, because the increase in precipitation occurs when that precipitation is predominantly snowfall. For now, anyway—at some point, the warming temperatures will overcome the increased moisture, and snowfall will trend downward even where it is currently steady or increasing.

We feel our climate not in averages, but rather in the daily weather and the impact that it has on us. We also see it in the changes in plant and animal life around us. For example, maple trees are a key

21. Analysis of data generated using the National Weather Service's Local Climatology Analysis Tool, available at https://lcat.nws.noaa.gov.

part of Wilder's stories about the Big Woods. The U.S. Forest Service projects that the "importance value," or dominance, of sugar maple trees relative to other species will decrease dramatically across all of Wisconsin by the end of the twenty-first century, as the tree species supported by future climates at any given location change.[22] Trees don't just pick themselves up and move; they must spread and adapt, and the rate at which they do is slower than the rate of our current, human-caused climate change. As a result, tree die-offs are likely in some tree species as climate changes, and that doesn't include the influence of invasive pests that may follow the changing climate. The risk of damage from pests such as the forest tent caterpillar and the pear thrip increases as the number of days above freezing in winter and early spring rises.

As we saw in the "sugar snow" episode from *Little House in the Big Woods*, the ideal temperature conditions for sugar maple sap runs occur in the spring when maximum temperatures at or above 40°F, with minimum temperatures below freezing, occur for a prolonged period of time.[23] Sugar maple trees that remain in the Pepin region of Wisconsin experience those ideal conditions for running sap about as often to more often in the late winter to early spring in recent decades, compared to Wilder's years there in the late 1800s. One of the impacts of climate change in western Wisconsin has been an increase in winter and early spring temperatures, with the potential for the ideal conditions for sap runs to shift earlier in the season. That said, hot weather causes maple tree sap to decrease its sugar content, so the additional hot days during the summer as the climate warms will mean that it takes more volume of sap to produce the same amount of maple sugar or syrup. The number of days with running sap has not changed much and is not projected to change much as climate changes, but trees that are too far south to maintain a tolerable climate will cease producing enough sap to collect. Climate projections suggest that sugar maples in Wisconsin

22. U.S.D.A. Forest Service Tree Atlas, accessed 6 May 2024, https://www.fs.usda.gov/nrs/atlas/tree/RFtreemod_318.html#.

23. Jerry Clark, "How Warm Weather Affects Sap Collection In Maple Trees," University of Wisconsin Extension, 28 Mar. 2016, https://www.wpr.org/agriculture/how-warm-weather-affects-sap-collection-maple-trees.

will be able to adapt to changing climate conditions, and the area will continue to support sugar maple growth, even if their dominance decreases.

Wilder noted in her 1919 article, "It seems such a pity that we can learn to value what we have only through the loss of it. Truly 'we never miss the water 'till the well runs dry.'" She went on, "But we have destroyed trees when it was not necessary, seemingly through a spirit of wantonness, and so we have a double task before us: to plant trees where they did not grow and to replant in some places where they have been cut down."[24]

We choose our ideal climate, and we also make choices that affect the climate around us, making it either more or less ideal. Decisions to plant or cut down trees directly affect the climate where we live, as does the decision to plant large tracts of land with thirsty corn and soybeans, to pave new roads and build new homes, and to use renewable energy sources like wind, water, and solar power.

Pilgrimages to the Wilder homesites are deeply meaningful to many, and yet they may not give us a faithful glimpse through Wilder's vantage. Visitors can see Lake Pepin and its pebbles, and the bluffs in which the Little House in the Big Woods was nestled. Young Laura said in the closing words of *Little House in the Big Woods*, "Now is now. It can never be a long time ago," but those days are indeed long ago, tucked away in a time, a place, and a climate that will never be replicated.[25]

Adapt or Move

As climate changes, some currently occupied corners of the country will likely become uninhabitable. In most cases, residents will be forced out by the ocean as it steals and drowns the land beneath them. Occasional flooding during peak high tides (known as "king tides") will become more common as ocean levels rise, then sneak into the everyday high tides. Pools will linger during low tides where the ground used to dry. The pools will get larger, connect to

24. Laura Ingalls Wilder, "The Wanton Destruction of Trees," *Missouri Ruralist*, Oct. 1919, in Hines, *Farm Journalist*, p. 69.

25. Wilder, *Little House in the Big Woods*, p. 238.

other pools, and become permanent waterways. In time, the pools will cover lawns, roads, driveways, and parking lots, squeezing out the humans who can no longer access their homes and businesses. The threat is especially apparent in flat coastal plains and barrier islands, where even a few inches of ocean levels rising can fill the low spots on land.

In other corners, the ocean is eating away the coastal land. Stormy seas churned up for longer seasons chew at the edges of the land like a child's teeth sinking into a solid chocolate Easter bunny. The land and anything humans built on it will drop into the ocean's throat, gone forever. The erosion is especially aggressive in places like Alaska, where sea ice used to blanket the ocean in the fall and calm it. As the protective cap of ice forms later and later, the storm systems typical of fall weather churn the seas for longer and longer storm seasons. One extra storm is just a nibble at the edge of the coast, but the repeated storms devour the land in chunks that can wipe out thousands of feet of land with each storm, each season. Entire communities that have thrived on Alaska's coast for thousands of years have fallen into the sea, swallowed, their humans left stranded and searching for a newer, safer home.[26]

Closer to the Ingalls and Wilder family homes, the changing climate shifts people from their homes less dramatically, though the effects are still tangible and real. Much like the coasts, one of the biggest drivers of climate shifts is water, though in a different form than it is in the coastal communities. The tug-of-war between drought and flood has its flag right over the central United States. On the western side of the Plains, Team Drought has its feet firmly dug into the cracked ground. While the dusters are not as dramatic as they were in the 1930s, thanks to the soil conservation measures enacted because of the Dust Bowl, they are not relics of the past. Dust storms and prairie fires still churn on the dry prairie winds as the drier climates of the Southwest creep toward the High Plains. The drying is spreading into the Rocky Mountains, where the drier

26. James W. Jordan, "Human and Economic Indicators—Shishmaref," NOAA Pacific Marine Environmental Laboratory, last updated Dec. 2006, https://www.pmel.noaa.gov/arctic-zone/detect/human-shishmaref.shtml.

climate and spreading pests have scorched vast tracts of pine trees. Even irrigated farming is becoming unsustainable on the western side of the Plains as groundwater levels fall farther below ground and demand from hot and dry weather increases at the surface.

To the northeast of the Plains, Team Flood firmly grasps the tug-of-war rope. Weather is getting wetter in the Great Lakes and upper Midwest, especially in the spring. The wet weather and wet ground have tempered some of the daytime heating trends with climate change, but nighttime low temperatures are warmer and more humid than ever, causing heat illnesses when heat waves wash over the area. Repeated floods are eroding riverbeds and damaging roads, bridges, and homes. The floods leach nutrients from the soil and carry them away from the hungry crops. Aging dams break and levees breach under repeated watery assaults.

As Team Drought and Team Flood pull and tug, the middle of the rope with its marker flag oscillates across the central Plains. The region gets the worst of both worlds: droughts on the nose of hot and dry weather spreading from the south and west, and floods washed in on wet and humid weather connected to the north and east. Farmers and ranchers hedge their bets and prepare in both wet and dry directions, absorbing the challenges added by both extremes. To some extent, this has always been the case in the central United States. With the changing climate, the extremes are magnified, and the range of possibilities is wider than it used to be. And the solutions are more expensive because of it.

For many farmers, ranchers, and landowners, the cost is too much to bear. Like the homesteaders of the past, including the Wilders, many will sell their prairie farmland and quit the business of trying to live off it.

As climate continues to change, some of today's farmers and ranchers will adapt to it and find a new balance, either because they can or because their financial situation means that they must. Some city dwellers will stay in their cities even as the heat makes them sick, because they lack the means to move elsewhere. Some suburban dwellers will find themselves in a newly expanded flood plain, paying the price through expensive, mandatory flood insurance. Some homeowners will find a way to absorb the new cost, while for others, the expense will be the last straw that forces them to move, whether they want to or not.

For many of us, where we live is a choice, and we will continue to make choices as the climate and other circumstances change around us. The same is not true for everyone, though. Some will be left to live in hostile environments, trapped by circumstances outside of their control. Those who stay will have to adapt to the changing world around them.

Laura and Almanzo Wilder found financial and community stability in Mansfield, carrying them through their middle-age years and to the sunset of their rich lives. It seems that they found relative weather and climate stability, too, smoothing out the rough edges of the climates of their past homes as they settled into the happy medium of the Interior Highlands of Missouri. At that time, in that place, the Wilders found the climate they needed not just to survive, but to thrive. They found their ideal climate.

To find an ideal climate now is a little more challenging. Climates are moving and shifting faster now than they did in Wilder's years, pushed by the hand of human-caused climate change. The ideal now may not be quite so ideal through the lifespan of a generation, or a mortgage, or a childhood. Those who want to follow an ideal climate will need to look ahead to where that climate might be in a few decades and then make some choices: Is it more important to be ideal now or later? What parts of the climate are changing the most, and are those the parts that make it more or less ideal for safety, health, or bank accounts? How will the changes threaten the safety and stability of a person, a family, or a community? How much will the changing climate cost in this location in added cooling expenses, flood insurance, storm damage repair, water access and use, recreation, and other expenses that are not yet clear?

The kinds of changes it takes to decrease how much the climate will change due to human causes can be costly, but it is just as true that not changing things also costs money. Climate change is expensive, and it threatens our safety by exposing us to more and stronger fires, floods, droughts, heat waves, and storms.

Climate change may force some people in the future to make

a difficult choice: stay where they are familiar with the land and community around them or move to follow a more ideal climate. The Wilders took a risk when they emigrated, not knowing if the new climate would suit them as well as they hoped or whether they would be able to make a living off the land that was so different from where either of them had grown up. Some farmers will decide that the land and its climate can no longer sustain the crops they've grown for generations, and they will make the choice between learning to grow something new or giving up the farms and changing their livelihoods.

Like Wilder, a few will write their stories for others to read. They'll tell their children, and a few of them will write it down. They will recall the climates of years past, a time unimaginable to future generations. They will become the storytellers for others, their voices adding to the chorus of Wilder and so many others.

12

Records, Stories, and the Future

To understand the weather of the future, we must first examine the weather of the past and present.

Think again about the record keepers: the storytellers like Laura Ingalls Wilder who weave weather and climate into their life stories, the farmers who take volunteer observations, the professionals who read carefully calibrated equipment, and the American Indian people who have drawn the world around them far longer than the settlers have kept scientific records. People from all walks of life have recorded weather in journals, spreadsheets, and web forms, noting everything from precipitation amounts to leaf emergence and robin returns.

Generations of weather watchers have provided windows into the weather of the past. Their trusted voices converge, like many instruments, into a symphony of weather past and present, the occasional discordant note resolved into resonance or outplayed by the tuned instruments around it.

Put another way, each observer contributes a piece to the large jigsaw puzzle of the climate picture. Any one piece is too small and limited in scope to complete the picture, but multiple pieces join together to fill in the gaps and reveal the bigger picture, even if some of the pieces have broken edges, tattered faces, or marker lines obscuring the picture.

Scientists do not need every detail of the picture to be present; they need enough to discern what is painted on the puzzle's face. More pieces and clearer pictures are welcomed and delightful, though, because every piece that fills a gap is one that might help someone else see the same picture that the scientist is seeing. It might provide a delightful new detail: a new color or brushstroke or object that only fits in that one puzzle piece but makes great sense in the larger picture.

Farmer's Eyes

We return to the farmer and longtime COOP observer, whom we last saw taking measurements of snowfall. Taking a swig of coffee from his aluminum travel mug, he walks back toward his barn and pauses to feel the warmth of the morning sun on his face as he scans his fields. It's spring now, the winter snow a distant memory displaced by spongy ground.

The warmth of the sun reminds the farmer that time is ticking. He must find time to prep the fields for planting between the copious spring rains. When his father ran the farm, planting came a couple of weeks later, giving a little more time for spring work. Rains seemed to fall in gentler showers and less often, leaving soil enriched instead of drenched. Everything feels like a race now—racing to prep his machinery; racing to beat the rain; racing to plant in the sweet spot of air temperatures, soil temperatures, and soil moisture; racing to make long-term plans for the changes that will eventually make his farm no longer viable for corn and soybeans.

The soggy ground beneath his boots is an unwelcome reminder of how wet the spring has been so far. It's normal to have rain in the spring, and it's even normal to get some swollen rivers and streams in some years as they drain away the snowmelt and the spring runoff. What feels unusual to this farmer is that it has been wet almost every spring recently. Planting windows have narrowed as the soil has been too soggy to run his machinery. Early planters worry about seed rot, even with the coatings and treatments on modern seed hybrids. Late planters feel behind on moisture through the rest of the growing season, worrying that the corn will still be in vulnerable reproductive stages when the inevitable hot and dry midsummer weather strikes.

Timing is everything when it comes to rain. Timing and rates. If the spring and summer's twenty-four inches of rain fall in periodic showers that soak into the ground, with respites between the showers, then the crops get a steady drink. If the rains fall in downpours of several inches at a time with hot and dry weather between, or if they all fall in April and May and leave the crop's roots steeping in puddles and shallow top layers of soil, then crop yields suffer.[1]

Nobody knows the air, the soil, and the land like farmers. They work the land with all five senses, understanding everything from the feel of the soil to the taste of the air to the shades of green and gold and tan washing over the rows of crops. They don't need a climatologist to tell them that things are changing, because they have watched the changes develop over generations of immersion in the air and land. This farmer has the weather instruments to put numbers on the changes he has seen, and he keeps track of his own observations, even though he knows that his readings are forever preserved in the NCEI's archives. He can run through his spreadsheets to find how much wetter the spring is now than it was a few decades ago, but that is mostly to confirm what he already senses.

It took a while for some of his peers to come around and admit that things were changing, that the weather of the past is not around anymore. Most of them still don't care to talk about why. The cause doesn't matter to them and digging into the "why" would just cause uncomfortable conversations at their gas station coffee stops and family dinners. What matters now is how it's going to affect their farms for the next generation—whether those farms will continue to operate, what technology they will need to do so, what they will be able to grow, and what they can do to balance when the rains come with what the crops need.

This farmer has been working with the university's cooperative extension office to try new water drainage techniques, hoping

1. Eugene S. Takle and William J. Gutowski, "Iowa's Agriculture is Losing its Goldilocks Climate," *Physics Today* 73, no. 2 (2020): 26–33, https://doi.org/10.1063/pt.3.4407.

to preserve nutrients in his soil and keep the topsoil on his fields instead of washing downstream. He's volunteered to test new hybrids. He doesn't use the words "sustainability" and "adaptation" and "mitigation" around his peers, but he applies the practices and keeps himself at the forefront of new ideas. He wants to be a model for his peers, and for his children and grandchildren, of making his farm last for future generations. His weather observations are not his only legacy.

Traditional Ecological Knowledge

North America's oldest line of storytellers and record keepers combine generations of understanding and awareness about their environment and ecology with the scientific research that supports what they have already perceived. National Climate Assessments weave traditional Indigenous knowledge into the body of evidence about climate in the United States, acknowledging its importance and value. Rather than creating counting calendar pictographs as their ancestors did, today American Indian people convey their knowledge in formats introduced by the colonizers' society: writing books, attending meetings, giving talks and testimonies, and publishing research, all while telling their stories.[2]

At a meeting in Anchorage, Alaska, in August 2004, scientists gathered in a small conference room to talk about the state's climate.[3] Many attendees were from the NWS, representing both local and na-

2. L. C. Jantarasami, R. Novak, R. Delgado, E. Marino, S. McNeeley, C. Narducci, J. Raymond-Yakoubian, L. Singletary, and K. Powys Whyte, "Tribes and Indigenous Peoples," in *Impacts, Risks, and Adaptation in the United States: Fourth National Climate Assessment,* vol. 2, eds. D. R. Reidmiller, C. W. Avery, D. R. Easterling, K. E. Kunkel, K. L. M. Lewis, T. K. Maycock, and B. C. Stewart (Washington, D.C.: U.S. Global Change Research Program, 2018) pp. 572–603, DOI:10.7930/NCA4.2018.CH15; K. Whyte, R. Novak, M. B. Laramie, N. G. Bruscato, D. M. David-Chavez, M. J. Dockry, M. K. Johnson, C. E. Jones Jr., and K. Leonard, "Ch. 16. Tribes and Indigenous Peoples," in *Fifth National Climate Assessment,* eds. A. R. Crimmins, C. W. Avery, D. R. Easterling, K. E. Kunkel, B. C. Stewart, and T. K. Maycock (Washington, D.C.: U.S. Global Change Research Program, 2023), https://doi.org/10.7930/NCA5.2023.CH16; Kimmerer, pp. 6, 41–43, 134–35.

3. Alaska Climate Users' Workshop, 24–26 Aug. 2004, Anchorage, Alaska, attended by the author.

tional interests. A few were from universities and research climate centers, and a few represented Alaskan tribes and communities. Government and research entities gave briefings about new ideas for weather observation platforms, new findings about the climate in Alaska, and new strategies for partnerships. Nearly two decades later, most of the meeting's attendees would be hard-pressed to remember even their own presentations, let alone the highlights from others. But one presenter shook the room, and he didn't need a single PowerPoint slide to make his point.

An Iñupiaq man from the community of Shishmaref, on the Seward Peninsula along the west coast of Alaska between Nome and Kotzebue, stood and faced the sea of non-natives. For hundreds of generations, his community—the Iñupiaqs, sometimes called the Alaska Inuit—thrived on subsistence living and artwork on a barrier island, relying on sleds by winter and boats by summer to reach the mainland before small planes added an additional means of transportation. In the span of just one generation—just a decade or two, really—the community that had thrived for a thousand years or more began crumbling into the sea. The waters of the Chukchi Sea (just north of the Bering Strait) freeze in the fall, sealing the sea before deep fall storms could whip up the waves. As the waters warmed, the freeze-up arrived later and later, and the fall storms that used to bring cold now brought relentless and hungry waves, devouring chunks of land away with each storm. The land, softened by permafrost melt under stark warming trends, melts into the ocean's mouth. Sea-level rise compounded the reach of the waves, like the child who disturbs the surface of a full bathtub rather than a half-full tub. In 2004, the man's village faced difficult decisions about whether to abandon their homesite of four hundred years, and when, and how.

By 2016, the village voted to move to the mainland, becoming one of the first communities in the United States to reach climate refugee status due to their need for immediate relocation.[4] Difficult questions remain about funding the relocation, establishing a new

4. Rachel M. Gregg, "Relocating the Native Village of Shishmaref, Alaska Due to Coastal Erosion," Climate Adaptation Knowledge Exchange, version 2.0. product of EcoAdapt's State of Adaptation Program, last modified 23 Oct.

site with new community services, and ensuring that the next location is sustainable for the generations to come.

Closer to Wilder's northern Great Plains, the Wind River Indian Reservation in west-central Wyoming is home to members of the Eastern Shoshone and Northern Arapaho tribes. Its members have observed the changes in mountain snowpack and water resources, temperatures, and weather patterns that affect their land, which is arid and prone to water shortages and drought. In particular, the mountain snowpack that provides their water has been decreasing through the years, melting earlier and providing less water than it did historically. The water shortages have caused steep decreases in the availability of water that residents use for irrigation and ranching, as well as the water that supports the fish and wildlife on the reservation. Working with scientists from climate support outlets such as the High Plains Regional Climate Center, the North Central Climate Science Center, and the National Drought Mitigation Center, the reservation's citizens have developed strategies to better observe, understand, and adapt to the changing climate around them.[5]

Tribal organizations are also actively engaged in increasing knowledge around both traditional and Western sciences, so that present and future generations can include climate in their planning and land stewardship. The Northern Cheyenne Tribe, with lands in southeastern Montana, offers apprenticeships for youth interested in bio-cultural restoration science, and their tribal leadership has created an Integrated Resource Management Planning exercise that includes climate change.[6] Ecological restoration ac-

---

2024, https://www.cakex.org/case-studies/relocating-native-village-shishmaref-alaska-due-coastal-erosion.

5. Crystal J. Stiles, Natalie A. Umphlett, and Mitch Cottenoir, "Building Adaptive Capacity in Tribal Communities of the Missouri River Basin to Manage Drought and Climate Extremes: A Case Study from the Wind River Indian Reservation," *Journal of Indigenous Research* 8, no. 3 (2020), https://doi.org/10.26077/x429-5d77.

6. R. T. Conant, D. Kluck, M. Anderson, A. Badger, B. M. Boustead, J. Derner, L. Farris, M. Hayes, B. Livneh, S. McNeeley, D. Peck, M. Shulski, and V. Small, "Northern Great Plains," in *Impacts, Risks, and Adaptation in the United States:*

tivities that incorporate traditional American Indian perspectives and culture have a better chance of success, in contrast to "pop-up restoration" activities introduced by outsiders that seem to be more about making the outsiders feel better than about leaving a healed and healthy environment.[7]

Just as Wilder's stories can combine with scientific observations to tell a more complete story about blizzards and tornadoes, the traditional ecological knowledge held by people across the country provides a more complete picture of weather and climate in their locations. Their language, histories, and oral and written stories fill in gaps about the climate itself, its changes through the years, and the effects of those changes on their livelihoods.

Weather Is Changing

Maybe one of these refrains sound familiar:

> "When I was a kid, we had snow every Christmas. Now it seems like we never get snow in December."
>
> "I used to be able to ice fish on this lake until April, but now it's not safe by mid-March most years."
>
> "We never had air-conditioning in this house when I was a kid, but after that last heat wave, I need to get a window unit, because we were so hot that we started to feel sick."
>
> "It never rains in July anymore."

Memories of the weather of our childhood—or even of five or ten years ago—are not always perfect. Wilder was much better than most folks at recalling specific weather from decades past. But often, people can remember things about those years past that just don't happen anymore, or things that never happened in the past

---

Fourth National Climate Assessment, vol. 2, eds. D. R. Reidmiller, C. W. Avery, D. R. Easterling, K. E. Kunkel, K. L. M. Lewis, T. K. Maycock, and B. C. Stewart (Washington, D.C.:U.S. Global Change Research Program, 2018) pp. 941–86, DOI:10.7930/NCA4.2018.CH22.

7. Jennifer Grenz and Helsey Geralda Armstrong, "Pop-up Restoration in Colonial Contexts: Applying an Indigenous Food Systems Lens to Ecological Restoration," *Frontiers in Sustainable Food Systems* 7 (2023), https://doi.org/10.3389/fsufs.2023.1244790.

that are happening now. Plants and trees flower at different times in the spring than before. Fishing for certain species peaks at a different time, or the fish have left familiar streams. Insects and pests that used to be a distant problem now invade neighboring yards or our own fields and forests.

Scientists can say with confidence that winters have warmed by two to four degrees from Wilder's years to ours, but those numbers don't mean much to most people. A warming of a few extra degrees in the winter sounds rather pleasant, perhaps taking the edge off the biting cold or reducing the heating bills. It isn't the average temperatures of months, seasons, and years that define how people experience the weather. It's those daily conditions, the big events and extremes, and the changes that the warming causes to the environment around us that become noticeable.

The winter of 2011–2012 never felt like it took a full grasp across most of the United States. Winter always will bring cold temperatures and snow to the northern half of the United States. But that winter, the cold came and went like the wind itself. Snow fell in fits and starts, an inch here, two inches there, with relatively warm spells between. The warmest day of December across parts of Nebraska and Iowa fell on New Year's Eve, a rare occurrence in a region that usually sees temperatures diving through the month. Brief warm-ups and light snow dustings continued into January. By February, the winter almost felt like winter, with more snow than in any other month. Temperatures were still above normal, though, with warm spells beginning right on 1 February. The February snow totals were below average in some parts of the Great Lakes and Midwest, while above average in others. Snows were still interspersed with warm-ups. On the last days of February, temperatures crept well above normal, and March came in like a lamb.

Throughout March, the warmth built up in swells like waves on a rising tide. Midwesterners kept waiting for the bottom to drop out, for the cold to return, as it inevitably does in March. The dissonance of the warmth hung in the air like a symphony holding a chord too long. Rather than the typical snows and the swings between warm and cold that the month usually brings, the temperatures stayed at May-like readings in the seventies and eighties nearly all month, ending in the nineties in Omaha. The last freeze in Omaha had been on 9 March, which would have been weeks early if it stuck. Confused spring plants emerged from the ground, their bright flowers sniffing the warm air with curiosity. Excited trees erupted with buds on their branches. March swept into April as it had entered—extremely warm. Old daily record high temperatures fell in swarms in March and April, and even monthly warmest temperatures broke in March.

The record earliest last-freeze dates in late March came and went. As early April ticked by, delicate fruit trees and garden flowers continued to emerge, fooled into thinking it was safe, becoming more vulnerable as each warm day passed. The forecasts hinted at a wobble, a break in the warmth. For just a night or two, under clear skies, calm winds, and dry air, the temperature threatened to dive below freezing one more time. NWS offices issued freeze warnings, but there is little that a fruit orchard can do to save hundreds of acres of emerging buds in the event of a freeze. Some growers may use bonfires, fans, or even helicopters to try to nudge temperatures just a couple of degrees warmer. Sometimes it is enough, if the orchardist is equipped to use these tools, but they only work under very specific conditions: the cold must be the kind that comes on a calm, clear night, known as "radiational cooling"; the temperature must be only a few degrees below freezing; the bonfires, fans, or helicopters must be able to reach enough of the trees in the orchard to be worth the expense of using them.

In the second week of April, the temperature dived below freezing one last time. In places like Nebraska and Iowa, the dip was brief and was only a degree or two below freezing, but still enough to burn the most sensitive of vegetation that had emerged—the peach trees and tulip blossoms. In the Great Lakes, the icy fingers of several nights of freezes devastated the cherry trees in Michigan, clawing away at least 90 percent of the tart cherry crop.[8] Vineyards and orchards suffered serious and expensive losses to

8. Saskia de Melker, "A Sour Season for Michigan's Cherry Farmers," *PBS NewsHour*, 16 Aug. 2012, https://www.pbs.org/newshour/science/science-july-dec12-michigancherry_08-15.

their fruits, with ripple effects as the lost fruits led to lower tourism and sales.[9]

The freezes of April 2012 technically came earlier than normal latest freezes, but the warmth that preceded them was so intense that the freeze was still devastating. Measured by calendar dates, the freeze was close to on time. Measured by warm-up and growth before it, the freeze was very late. The spring of 2012 provides a stark example of hazards in a warming world. The rash of high temperatures in early March from the Great Lakes to the Corn Belt meant that buds and leaves emerged far too early in the season. When the inevitable freeze returned, nipping those buds and leaves, it prevented the trees from bearing fruit. The freeze ended up costing millions of dollars in damage and lost revenue.[10] The resulting scarcity drove up the price of fruit across the United States, spreading the effects of a regional freeze to the wallets of the entire country.

A similar situation in 2007 produced another billion-dollar disaster across eighteen states from the Plains to the Southeast. After a warm March, freezing temperatures returned during the week of 4–10 April, devastating fruit trees, nursery plants, and some early-season crops.[11] These freeze events used to be decades apart, and now they have been occurring more frequently as temperatures rise in late winter and early spring but on-time freezes still occur.

From the first robin of spring to the emergence of lilac flowers, well-known plants and animals are visual markers of the environment around us. The National Phenology Network tracks plant emergences, bird migrations, insects, and more across the United States, allowing citizens to add their observations to their database to fill in gaps. The Audubon Society also gathers data from back-

9. Rosemary Parker, "Loss of Michigan apples, cherries, peaches to freeze has far-reaching consequences," MLive, 19 May 2012, https://www.mlive.com/news/kalamazoo/2012/05/no_michigan_apples_cherries_pe.html.

10. Melker, "A Sour Season."

11. *The Easter Freeze of April 2007: A Climatological Perspective and Assessment of Impacts and Services* (NOAA/USDA Technical Report 2008-01, 2008), accessed 8 May 2024, https://www.ncei.noaa.gov/monitoring-content/billions/reports/20070404-20070410-freeze/tech-report-200801.pdf.

yard bird watchers to help track bird migration and populations. As the citizen scientists collect their reports through the years, the professional scientists can track them for patterns and trends. Where are the plants budding earlier or later than they used to in the past? Where are birds and insects not returning at all anymore?

Scientists combine the things that people feel like they are seeing—changes in weather, wildlife, plants, and pests—with the data, working to determine whether the perceptions are supported by evidence or based on a skeleton that is too incomplete to make out the shape of the animal.

Beyond the Shadow of a Doubt

In an unassuming office painted yellowish-beige, just big enough for a simple laminate corner desk and matching tall bookshelf stuffed with textbooks, conference binders, and sheaths of printed research papers, a scientist adjusts her glasses in her standard-issue office chair. In pull-on black pants and a bright short-sleeve printed shirt, with the requisite office cardigan on top against the chill of too much air-conditioning, she scans tables of numbers on her dual-monitor computer display. A powerful computer tower whines under the desk, churning through code that converts math and physics equations into temperatures, precipitation, and weather variables decades in the future. Next to the monitors, her laptop chimes with incoming chats and emails from colleagues, students, and family, all seeking constant contact and quick answers. The easy questions ("What's for dinner?" and "What's the deadline for this paper?") are easy to quell with a few keystrokes, but in front of her, the more difficult questions tangle in the results of her computer's output: What will the climate at this location be in fifty years? How much did climate change contribute to the most recent weather disaster?

The scientific process is often sparked by questions that follow questions. Answers are rarely straightforward in the scientific world; they come with footnotes and asterisks, most-likely and almost-but-not-quite-impossible scenarios. Questions like "What happened?" are often followed by the much more complicated "Why?" and "How?" and the inevitable "Will it happen again, and when?" A weather event like a heat wave, or a seasonal event like an unusually warm winter, drives scientists to start down the path of asking questions. The research might begin with a more straightforward analysis of the case in question, collecting evidence to make a compelling argument for what happened. Just as in a court case, although this is the most straightforward part of making a case, it can still end with gaps in the evidence.

The next layer of argumentation adds the complex argument of motive, the "why" of the event. Did an unusual global weather pattern contribute to the heat, or warm ocean temperatures, or a warmer background climate as temperatures rise around the globe, or the random and chaotic nature of weather itself? There is rarely one answer. Climate change underlies all weather. It simply is a question of whether scientists using current technology and understanding can detect that signal among all the others.

New scientific concepts, theses, and findings are subjected to a process called peer review, in which researchers write papers to share and support their findings. Scientists discuss results with their peers in both formal and informal settings. As colleagues explore these cases and events, they add to the body of knowledge. Sometimes, the scientists' results converge into concepts that become a fundamental part of the science. Sometimes, the scientists produce several explanations to the question of "Why?" and leave open the potential for further study, limited by current understanding of the science, available observations of the event in question, or computing power to replicate it. An explanation holds up when no credible alternative explanations can survive the scrutiny of others in their field of study.

Consensus among scientists is a rare gem, achieved through years and even decades of compelling research from multiple people using a range of approaches to come to the same conclusions. When scientists reach that consensus, confidence is high in the principle, even if some of the small details surrounding it are still under investigation. New findings nudge the course of the existing body of work, but our knowledge has come so far that there is rarely any watershed discovery that turns the course of an entire field of science on a dime. The fundamental concepts are the ones now accepted beyond the shadow of a doubt, with the emerging science building knowledge by a preponderance of evidence.

Scientists have been studying the effects of human activities on the climate for centuries. Fundamental work on understanding how burning fossil fuels releases carbon dioxide into the atmosphere, how carbon dioxide turns the knobs on the Earth's temperatures, and how more carbon dioxide leads to more warming has been documented in scientific research papers since at least the mid-1800s. None of this information is new or controversial. Eunice Foote proposed the relationship between carbon dioxide and atmospheric warming in 1859, or around the same time that Almanzo Wilder was born (depending on the accuracy of his birthdate, which is indeed controversial).[12] Some of the tuning knobs are working on understanding some of the details, such as how clouds interplay with temperatures in a warmer planet and how zones of severe thunderstorm, tornado, and hurricane activity will change. Scientists are actively working on many of the details surrounding climate change. But the concept that human activity changes the climate is as fundamental and strong in science as the concept of gravitational pull.

Experts agree: human activities are a stronger contributor to climate change than anything else (even the solar and orbital cycles, continental drift, and volcanoes). Generations of climate research have determined that there is no other realistic explanation for the changes to weather and ocean patterns that have been observed.

It's mostly bad. Climate change causes problems because humans evolved and adapted and built to a climate that is slipping through our fingers, and our food, homes, and sources of water are in danger, not to mention the ecosystems that support and weave together with human life.

There's hope. It's not too late to fix it, but the longer humans wait to act, the steeper the cost will be in both dollars and lives.[13]

12. NOAA, "Five Significant Moments in Climate Science History," 17 Jan. 2024, https://www.noaa.gov/heritage/stories/five-significant-moments-in-climate-science-history.

13. Ed Maibach's "Six Truths" are the following: It's real. It's us. Experts agree. Others care. It's bad. There's hope. *See* Nicholas Badullovich, John Kotcher, Te-

Nothing New Under the Sun

Imagine a world where more people live like Laura Ingalls Wilder did.

Wilder's column in the *Missouri Ruralist*, "As a Farm Woman Thinks," was packed with tips about growing her own food (both plant and animal sources), fixing and reusing her own things, and using materials from her own land. Her local fame for raising chickens was part of the reason the *Missouri Ruralist* invited her to write for the paper, and she used the chickens for both income and her own food source. Indeed, learning from past failures, she and Almanzo had diversified their income streams at Rocky Ridge Farm, and besides chickens the Wilders kept dairy cows and grew corn, wheat, oats, apples, strawberries, grapes, and a full vegetable garden to feed their family. Proudly, Wilder wrote in one of her first published *Missouri Ruralist* articles: "Our little Rocky Ridge Farm has supplied everything necessary for a good living and given us good interest on all the money invested every year since the first two."[14]

Wilder also wrote of the joys of dreaming up plans for her garden in the winter: "In imagination we see the plants in our spring garden, all in straight, thrifty rows with the fruits of each plant and vine numerous and beautiful as the pictures before us."[15] As with the livestock they raised, some of their crops went to market, and much went to feeding the livestock.

The Wilders famously built their farmhouse from materials on their property, customized to their needs. Rose Wilder Lane wrote in her closing chapter to Wilder's published journal *On the Way Home*, "It would be a white house, [my mother] said, all built

---

resa A. Myers, Seth Rosenthal, Anthony Leiserowitz, and Ed Maibach, "Understanding Six 'Key Truths' about Climate Change Predicts Policy Support, Discussion, and Political Advocacy," *Climatic Change* 178, no. 89 (2025), https://doi.org/10.1007/s10584-025-03934-3.

14. Laura Ingalls Wilder, "Rocky Ridge Farm," *Missouri Ruralist*, July 1911, in Hines, *Farm Journalist*, p. 39.

15. Laura Ingalls Wilder, "Make Your Garden!," *Missouri Ruralist*, Feb. 1918, in Hines, *Farm Journalist*, p. 85.

from our farm. Everything we needed to build it was on the land: good oak beams and boards, stones for the foundation and the fireplace."[16]

Wilder saw her farm home as a model for others:

> I would like to see our farm homes built, not for the present generation only but for our children and our children's children. . . . I have a fancy that the farm home should seem to be a product of the soil where it is reared, a permanent growth, as it were, of conditions surrounding it wherever this is possible; and nothing gives this effect more than a house built from rocks from the fields. Such a house, well built, will last for generations. Cement is another material of which lasting and beautiful buildings may be made. Even the common earth, the soil beneath our feet, can be used as building material and will last for hundreds of years. Tamped earth is one of the very oldest of building materials. In New Mexico and Arizona are walls made of it that are 4,000 years old, and it is still being used in various parts of the world.[17]

Wilder even commented on the rising prices of her usual goods at the stores that stemmed from waste instead of quantity or quality: "I am sure that one thing causing us a great deal of trouble and making much higher the cost of living is the extra price we pay for fancy packages."[18] While some packaging certainly keeps food and goods safer, anyone who has battled a bulging plastic blister package to extract a child's toy or piece of hardware knows that it is frequently excessive.

Wilder embodied the principles of "sustainability" and "buying local" long before such ideas entered the lexicon. She was a careful steward of her lands, writing, "It should be a matter of pride to keep our own farm, that little bit of the earth's surface for which we are

16. Wilder, *Little House Traveler*, p. 112.

17. Laura Ingalls Wilder, "My Dream House," *Missouri Ruralist*, Mar. 1920, in Hines, *Farm Journalist*, pp. 34, 69.

18. Laura Ingalls Wilder, "Income and Expenses," *Missouri Ruralist*, Oct. 1920, in Hines, *Farm Journalist*, p. 78.

responsible, in good condition, passing it on to our successor better than we found it."[19]

As important as stewardship of her land was to Wilder, so was stewardship of her finances. It was imperative to find ways to conserve money. Many others found themselves in a similar lot, especially during the Great Depression of the 1930s. Families didn't need slogans like "Reduce, Reuse, Recycle" to know that they'd save precious pennies by using less, reusing things they had, and making something new out of something old. Children of the Depression became the parents of Baby Boomers, grandparents of Generation X and Millennials. Even as a disposable economy exploded in the later decades of the twentieth century, many will remember their grandparents washing and reusing aluminum foil and plastic bread bags, cutting old shirts into fabric squares that became quilt blocks, and patching up pants.

Imagine the impact an entire generation could have today if it collectively decided to repair and reuse its clothing for a decade, rather than buying a new wardrobe every season. Imagine the impact of a collective community effort to refill reusable water bottles at community drinking fountains and spigots, instead of buying water in plastic bottles (Wilder's "fancy packages"). Imagine building houses made to last for generations, using materials that could withstand the elements where they were built, using local stones to set its walls and foundations, as the buildings and dwellings merge into the color palette of the surrounding countryside. Imagine crafting quilts from remnants of tattered clothing, then snuggling under them to read together on cold winter nights—screens off, just for a while, and in-home heating systems turned down just a notch or two to balance the heat trapped beneath the homemade quilt. Imagine building more furniture from the trees and materials near us or salvaging usable pieces of furniture to re-shape into the next heirloom.

With her words, Wilder has left a blueprint for the path to living as a good steward of land and home. She wrote of recipes that stem from affordable ingredients (reinforced by Barbara Walker's

19. Laura Ingalls Wilder, "Heirs of the Ages," *Missouri Ruralist*, Nov. 1923, in Hines, *Farm Journalist*, p. 47.

The Little House Cookbook). She described the construction of entire homes and many furnishings when the family lacked even nails to hold them together. She patiently stitched the stories of sewing necessities and garments, even if she hated sewing. Wilder left a legacy of stewardship for those who are willing to put in a little work to follow in her footsteps.

Auld Lang Syne

What weather events from your childhood stand out in your memory? Was it a big storm that tore shingles off the roof and branches off the trees? Was it a winter storm so cold that the cars wouldn't start, the roads were snowed in, and schools were closed for a week? Was it a heat wave, back before you had air-conditioning, when you lay awake at night with a fan just blowing the hot air around in circles and a cool washcloth on your neck, leaving your sheets damp with both water and sweat?

What is it about that weather event that you remember? It may have involved nearly all your senses. It could be the feel of searing heat or blistering cold on your skin, the sting of rain or snow or sleet, the scorched air of a relentless drought. It could be the smells of acrid smoke, rain on dry ground, freshly broken trees, salty water flooding far inland. It could be the sounds of crashing thunder, roaring winds, hammering hail, floodwaters rushing like a waterfall, the muffled stillness of a calm and deep snowfall. It could be the bright whiteness of a snow-covered landscape, the gloomy gray of everlasting clouds, the brown of a landscape that should be green, landmarks destroyed, brave sunshine filtered by layers of smoke, rainbows, sunsets, and clouds of all shapes from comical to threatening.

Maybe it was a singular event: a quick shot like a storm that passed in mere minutes, or a snowstorm that started in the evening and finished in the morning. Maybe it was a season: the snowiest and coldest winter you can remember, or the spring when rains would not let up and the rivers swelled. Maybe it was an amorphous memory that blankets your childhood: ice skating on the lake, reaching the sandbar when the tide was out, harvesting a garden in the chill of a fall morning.

Wilder must have felt relief waking on wintry mornings in Mansfield, Missouri, to know that the biting cold, ceaseless wind, and insipid blizzards of her childhood could not stretch their icy claws into Missouri without the edges being filed dull by the distance. After she moved south, her childhood memories of winter or any of the other weather she experienced provided no basis for comparison to the winters of her middle and elder ages.

People hold in their hearts the weather that framed their lives, their past, their histories: the snows that used to cover a favorite sledding hill, the ice that used to hold long enough to support the weight of an ice shanty into the early spring, the range of the swings between wetter and drier years. They can hold those years in their memories and also take notice of what they see and feel today. The robins arrive earlier, and the geese fly south later. Flowering trees are tricked into budding earlier and often get nipped by a freeze for their trouble. Dry spells get drier and hotter, wildfire seasons grow longer, and swaths of trees wither. People see the changes in the weather around them, illuminated by the alterations in plants, animals, rivers, and soils they have known all their lives.

Those who live closest to the land and make their living from it often have a deep understanding of the relationship between the land and the weather, honed by a lifetime of experience. Wilder herself recognized that humans changed the land and air around them, just as the folks who work the land and waters today see.

Wilder's stories have touched generations of readers with the vivid details of her childhood. Using all her senses and digging deep into her memories, she gave readers a view into her world, through her eyes and ears, fingertips, nose, and tongue. She paid attention to the world around her, particularly after her sister Mary became blind and needed Laura's descriptive words as a substitute for her sense of sight. Wilder collected the stories that shaped her values and moved her heart, created heroic figures out of her parents, and grounded her storytelling in the very real memories and events of her youth. Then, she sprinkled in a little extra storytelling to make her stories even more pleasing to her readers.

Wilder's stories alone cannot provide a full record of the weather and climate everywhere she lived, but they tell part of the story of living in those climates with those weather conditions. She did not measure snow depth after every blizzard, but she created the chill and disorienting winds in her stories and then drew her readers to see the snow levels through her eyes, from a second-story window vantage or a peek into a snowy tunnel out of the lean-to. While record keepers cannot build a complete picture of the weather of Wilder's days from only her stories, they can pair her stories with the measured temperatures and precipitation to deepen the understanding of those weather events of the past that people now can never see again. We especially trust the picture of past weather when the stories like Wilder's converge with the records, and the "birds in their little nests agree."[20] When they don't match, the mystery invites deeper digging to explain the differences.

The connections go beyond stitching together the weather records and Wilder's stories, though. The weather around us affects not just the people who live somewhere, but also the soil, waterways, plants, and animals around them. They can be immediately impacted by weather, such as a crushing ice storm that snaps branches from standing trees, or they can be impacted over time, such as a trend toward drier summers that makes a once thriving stand of trees desiccate and wither.

From technology to society to climate, progress and change happen around us, even if we stay in the same place ourselves. The progress and change of the world moves like a river flowing through our lives. Some folks prefer to stay in that place, a rock in the stream watching the waters and maybe trying to shape them a little as they rush by. Others prefer to float in the waters, following the current as it flows. Maybe they will float on with that river all the way to the ocean, riding the water cycle to begin the journey again. Maybe they will land on a bank and stay for a while, enjoying rest in a place of comfort before they move on. The rocks in the stream, the perpetual-motion particles, and the intermittent floaters all shape the character of the river itself. They're all a part of the same system, even if they view the river from different perspectives.

20. Wilder, *Little Town on the Prairie*, p. 152.

We can reminisce about the past, recalling it fondly with nostalgia and reverence. At the same time, things will never be like they were in the past. The waters move downstream, the riverbed shifts, and the changes happen whether we want them to or not. Whether we move with them or they move around us, changes happen as surely as a river flows. And if people want to navigate that river, they need the most updated navigational map to do it safely.

Likewise, if people are going to live in the climate of today and the future, they need to make sure they have the most updated map of it so that they can do so safely. It would be a waste of money and effort to prepare for a summer with more than twenty inches of rain occurring only once every generation or so when the rain now exceeds twenty inches nearly every year. People can compare the now and the future to the past, but the wisest use of time and money is to prepare for the future, rather than preparing for a past that no longer exists or assuming that the weather will stay the same as it is now.

Science gives information about the direction of the future, just like a map gives information about the direction a river flows. We don't have to bump along the banks, landing and launching and wondering where the next voyage will take us. A map might not contain every detail about the journey. Just the same, it will tell us what direction we can expect to go if we stay in this stream. We can understand where the waters are rushing as they flow past the rock in the stream.

And like Wilder, we can pay attention. We can tell the stories of weather in our past and present. We can become part of the narrative.

The greatest gifts of the past are the lessons taught to future generations. How will the storytellers of tomorrow convey today's world? What fossils will scientists in the future unearth about weather systems now? Scientists and storytellers honor and remember the past by writing about it, studying it, and learning from it, carrying those lessons forward to the present and future.

Epilogue

The story of weather and climate does not end here. You and I and many others continue to be part of the perpetual narrative. Laura Ingalls Wilder's piece of the story continued until her last letters, bequeathing the work of paying attention to the weather to the next generations. In 1950, she wrote, "The memories of our childhood are so much alike and so different from now, they should make us friends. What a thrill we had on some sunny spring day when we first took off our shoes and stockings and ran outdoors barefoot for a short time."[1] You and I pick up the narrative in our lifetimes and pass it along to our children after us.

It's true that trust in stories of past weather grows when the data and observations agree, that is, when the bones underneath support the details added by story. The reverse is also true: trust in the data and observations of past weather also grows when supported by the human stories and experiences that fill it out. The tragic loss of Mary Ingalls's vision prompted Wilder to take closer notice of the world around her. In *By the Shores of Silver Lake,* Pa's directive to Laura required that she put her active imagination into words and follow in his footsteps as a storyteller. "On that dreadful morning when Mary could not see even sunshine full in her eyes, Pa had said that Laura must see for her. He had said, 'Your two eyes are quick enough, and your tongue, if you will use them for Mary. And Laura had

1. Laura Ingalls Wilder to Miss Panzram, 3 June 1950, in *The Selected Letters of Laura Ingalls Wilder*, ed. William T. Anderson (New York: Harper, 2016), pp. 324–25.

promised."[2] The voices of past observers, scientists, and storytellers like Wilder lift together like a choir, resonating in their harmony of past weather.

The professional scientists who both record and interpret past weather have interests and motivations similar to yours. They too are humans with families and cherished loved ones, neighborhoods, and communities they want to protect. They spend lifetimes seeking not just answers, but more questions, in their goal to find explanations to mysteries that hold up to the scrutiny of peers and time.

The surest way to protect people and communities from wilder weather is to pay attention to it. Look up at the sky and look around you. Notice what parts of your life, your community, and your surroundings are most impacted by the different types of weather you experience. Maybe even take some notes, jot your observations in a journal, or draw a picture. Teach your children to pay attention, too. Voices like yours are the ones that will give color and texture and sound to the fossilized bones of weather when others look back at your weather events in generations to come.

If paying attention in an organized way appeals to you, there are several outlets. The Community Collaborative Rain, Hail, and Snow (CoCoRaHS) network takes observations from volunteers, all using the same simple plastic gauge, to gather precipitation readings across the country.[3] CoCoRaHS observations are archived and used in scientific studies, crossing that bridge from the flesh of the weather around you to the bones that will be fossilized for years to come. If watching plants and animals is more appealing, the National Phenology Network collects volunteer observations of animal sightings and plant emergence to track their progress seasonally across the United States.[4] If your interests are specific to birds, you can provide your reports to the National Audubon Society.[5] If you are drawn to storms, you can take training from your nearest Na-

2. Laura Ingalls Wilder, *By the Shores of Silver Lake*, pp. 22–23.

3. CoCoRaHS is available online at https://www.cocorahs.org.

4. National Phenology Network is available online at https://www.usanpn.org/nn.

5. Information on National Audubon Society local chapters is available online at https://www.audubon.org/about/audubon-near-you.

tional Weather Service office to become a Skywarn spotter, reporting thunderstorm winds, hail, and tornadoes from a safe location and directly helping National Weather Service meteorologists keep your community safer with their severe thunderstorm and tornado warnings.[6] These reports, too, become part of the bones, archived by the National Weather Service and the National Centers for Environmental Information for generations ahead to look back and study.[7]

As you pay closer attention, you will notice the ways that weather and climate affect you, your loved ones, and your surroundings. Many natural disasters are intertwined—weather events, occurring in primed and vulnerable environments, affecting people in harm's way. A forest fire, sparked by lightning, ignites and accelerates rapidly among trees desiccated by drought and pest infestations, spreading into a community built in a burn-prone mountainside. A hurricane pushes a storm surge on top of rising sea levels, flooding a community where people live in poverty and cannot afford the cost of evacuation. A hot and dry summer week extends for a month, creating flash drought that withers corn on the stalks and evaporates farm ponds.

When rare weather events become more common, they highlight the connection between those weather events and the climate that drives them. Just one rare event can happen by chance; when they repeat, it's a warning sign of what the weather could be in the climate of the future.

People naturally connect with the weather of their pasts and even the weather in the past of someone they recognize, like Laura Ingalls Wilder. Nostalgia for what used to be also serves as a reminder that the weather of the past is gone, evaporated on the winds. Someday, the weather today will be the past that the next generations remember with nostalgia, perhaps wishing for those days as they struggle with the weather to come in a future climate.

The weather and climate around Wilder and her family affected

6. Learn more about National Weather Service storm spotter, or Skywarn, training online at https://www.weather.gov/skywarn/.

7. NCEI Storm Events Database available online at https://www.ncdc.noaa.gov/stormevents/.

everything from their daily lives to big decisions to move and start something new. Even as Wilder's readers learn about it, they know that her world looked and felt different from the world now. The human fingerprint was not as visible or sensible. More of the vegetation was native, and fewer houses and buildings and roads crisscrossed the United States. It was a cooler world, and in some parts of the United States, a drier or wetter one as well. At that time, Euro-American settlers had not observed the weather conditions for long, and they did not yet understand their full possible range.

Even with modern knowledge and technology, we remain vulnerable to the weather. Look no further than traffic jams on snow-covered interstates, homes and businesses invaded by floodwaters, and stunned faces after a tornado ravages a community. Even when we can see it coming, we are vulnerable to the impacts. We may be at risk in different ways than the Ingalls and Wilder families were, but like them, we can lose our homes, be forced to evacuate or emigrate, or suffer injury or illness as a direct or indirect result of weather or climate.

Both weather and climate can cause anxiety and trauma. Weather anxiety and phobias can stem from an innate fear of weather extremes or the trauma of experiencing a weather disaster (or near-miss).[8] Thunderstorms, tornadoes, gusty winds, and hurricanes are common triggers for anxiety and fear, but so are heavy snow, ice storms, and flash floods. Often, the anxiety or phobia will arise when a weather forecast mentions the triggering hazard or when the conditions feel like they did during a past weather disaster (such as a humid spring day with a gusty wind). Climate anxiety, by contrast, is a pervasive feeling of hopelessness, despair, or doom caused by impending harm due to climate change. The anxiety can be caused by fear or worry about threats to people's livelihoods, concern for the well-being of future generations, despondence about the lack of movement toward solutions, and hopelessness

8. Jill S. M. Coleman, Kaylee D. Newby, Karen D. Multon, and Cynthia L. Taylor, "Weathering the Storm: Revisiting Severe-Weather Phobia," *Bulletin of the American Meteorological Society* 95, no. 8 (2014): 1179–83, https://doi.org/10.1175/BAMS-D-13-00137.1.

about the scale of potential harm.[9] In both cases, the anxiety can be alleviated by understanding and becoming part of the solution. People with weather anxiety can ensure they have emergency kits and preparedness plans for the weather hazard that triggers them, as well as reliable sources of information. Climate anxiety can be channeled into personal or community action to help mitigate or prepare for changing conditions.[10]

We are lucky, in these modern times, to have access to weather forecasts that help us plan our days and find safety before dangerous weather strikes. We have long records of weather observations to help us understand the range of past weather conditions. We have climate outlooks that can help us anticipate how the range of possibilities may lean one way or another in a coming season. The farmers, ranchers, settlers, and inhabitants of years past would have thought themselves very fortunate to have such insight that we have at our fingertips on radio, television, internet, and phone applications.

In the changing climate around us, the weather of the past might not always serve as a benchmark for the weather we will encounter in our future. Some events may surprise us, just as they surprised Wilder and her family, impacting our homes, our communities, and our loved ones. Science will help us understand what is possible, and technology may help us mitigate those impacts, as we, too, are explorers in new climates as they come to us.

Most of all, I invite you to talk to each other about the weather of now, and the weather of the past, and the weather of the future. It isn't so scary to talk about climate: it's just all of the past weather collected, the weather of now and all the forces that move it, and the weather of the future. Our conversations, friend to friend and

9. Catriona Soutar and Anne P. F. Wand, "Understanding the Spectrum of Anxiety Responses to Climate Change: A Systematic Review of the Qualitative Literature," *International Journal of Environmental Research and Public Health* 19, no. 2 (2022): 990, https://doi.org/10.3390/ijerph19020990.

10. Thijs Bouman, Mark Verschoor, Casper J. Albers, Gisela Böhm, Stephen D. Fisher, Wouter Poortinga, Lorraine Whitmarsh, Linda Steg, "When Worry About Climate Change Leads to Climate Action: How Values, Worry and Personal Responsibility Relate to Various Climate Actions," *Global Environmental Change* 62 (2020), https://doi.org/10.1016/j.gloenvcha.2020.102061.

neighbor to neighbor, will connect us with each other, and when we most need to come together to protect our families, loved ones, communities, and surroundings, we will already have established the relationships we need to work together and protect what we cherish.

Appendix 1
Ingalls and Wilder Family Weather Timeline

- 13 February 1857 or 1859: Almanzo James Wilder is born on his family's farm four miles east-northeast of Malone, New York.
- 4 July 1859: A very late freeze strikes Malone, New York.
- 7 February 1867: Laura Elizabeth Ingalls is born in Pepin, Wisconsin.
- 1868: The Ingalls family leaves Pepin, Wisconsin, initially bound for Chariton County, Missouri.
- September 1869: The Ingalls family arrives and settles on land belonging to the Osage Diminished Reservation, twelve miles southwest of Independence, Kansas.
- Spring 1871: The Ingalls family leaves their cabin near Independence, Kansas, and returns to Pepin, Wisconsin.
- October 1871: Wildfires destroy swaths of forest across the Great Lakes. 8 October 1871 sees major fires in Chicago ("The Great Chicago Fire"), Peshtigo, Wisconsin, and lower Michigan.
- Late March/early April 1872: A "sugar snow" probably falls in and around Pepin, Wisconsin.
- April 1874: The Ingalls family likely crosses the Mississippi River at Pepin just ahead of river ice break-up on 22 April as they leave Pepin, Wisconsin.
- 1874: The Ingalls family arrives near Walnut Grove, Minnesota.
- 12 June 1874: Rocky Mountain locusts arrive in southwestern Minnesota.
- 1874–1875: The "other hard winter" impacts the upper Midwest, including southwestern Minnesota, bringing extreme cold and above-normal snow.
- 1875: James Wilder family (Almanzo's family of origin) moves from Malone, New York, to Spring Valley, Minnesota.
- 1875: Rocky Mountain locusts return to Minnesota, invading Walnut Grove.

- 7 July 1876: The Ingalls family leaves Walnut Grove, Minnesota, bound for Burr Oak, Iowa, after a stopover with the Peter Ingalls family near Zumbro Falls, Minnesota.
- September 1876: The Ingalls family endures a "cold miserable" journey from Zumbro Falls, Minnesota, and arrives in Burr Oak, Iowa.
- August 1877: Rocky Mountain locusts depart Minnesota.
- August or September 1877: The Ingalls family departs Burr Oak, Iowa, and returns to Walnut Grove, Minnesota, living in town.
- Spring 1879: Almanzo Wilder files a homestead claim near De Smet, Dakota Territory (South Dakota).
- 9 September 1879: The Ingalls family departs Walnut Grove, Minnesota, to follow railroad construction into eastern Dakota Territory (South Dakota). They stay in the surveyor's cabin near De Smet, South Dakota, for the winter of 1879–1880.
- 19 February 1880: Charles Ingalls files a homestead claim 1.5 miles southeast of De Smet, Dakota Territory (South Dakota). The Ingalls family moves to the claim about a month later.
- October 1880–April 1881: The Hard Winter devastates much of the United States, from the northern Rockies to the Great Lakes and south to the southern Plains. De Smet is in the heart of the most severe conditions.
  - 15–18 October 1880: The October Blizzard.
  - 3–4 January 1881: The Chicago and North Western railroad company ceases train operations to De Smet.
  - 16 February 1881: Almanzo and Cap (presumably) take the Seed Wheat trip.
  - March–April 1881: Major river flooding from melting and runoff of Hard Winter snow.
  - 27–30 March 1881: Ice jam flooding causes significant damage to towns along the Missouri River, including Vermillion, South Dakota, and Green River, Nebraska.
  - 13 April 1881: The chinook arrives, marking the end of the Hard Winter; temperatures rise above freezing for the rest of the spring.

- 4 or 5 May 1881: The first train after the Hard Winter passes through De Smet.

- 4 January 1884: Laura and Almanzo endure the "Cold Ride" to De Smet from the Bouchie residence.
- 21 July 1884: Storm in De Smet.
- 28 August 1884: A tornado outbreak strikes eastern Dakota Territory (South Dakota), including the tornado near Howard, Dakota Territory (South Dakota), that is both observed by Laura and Almanzo and photographed by F. N. Robinson, making it the second-oldest-known photograph of a tornado.
- 25 August 1885: Laura Ingalls marries Almanzo Wilder, and they move to Almanzo's tree claim two miles north of De Smet, Dakota Territory (South Dakota).
- 5 December 1886: Rose Wilder is born on the Wilders' homestead near De Smet, Dakota Territory (South Dakota).
- 1887–1896: Droughts devastate crops and ranching in Dakota Territory (South Dakota), leading to an exodus of population.
- 12 January 1888: The "Children's Blizzard" strikes Dakota Territory, Nebraska, and Minnesota.
- 2 April 1889: A windy dust storm and prairie fire threatens the Wilders' homestead.
- 1890: The Wilder family leaves De Smet, South Dakota, bound for the home of Almanzo's parents in Spring Valley, Minnesota.
- 1891: The Wilder family leaves Spring Valley, Minnesota, bound for Westville, Florida.
- August 1892: The Wilder family leaves Westville, Florida, and returns to De Smet, South Dakota.
- 17 July 1894: The Wilder family leaves De Smet, South Dakota, bound for Mansfield, Missouri.
- 30 August 1894: The Wilder family arrives in Mansfield, Missouri. They move onto the Rocky Ridge Farm near town, where they live for the rest of their lives (minus a year in town).
- 1930–1939: The Dust Bowl brings drought, dust storms, and heat to much of the central United States, causing catastrophic crop loss, financial hardships (exacerbating the Great Depression), and physical and mental health tolls.

- 21 September 1938: The Great New England Hurricane of 1938, or the "Labor Day Express," devastates the Long Island to New England region, while Rose lives in Danbury, Connecticut.
- 12 April 1945: An F3 tornado strikes a path from Bradleyville to Mansfield, Missouri, knocking down trees on Rocky Ridge Farm. Its thirty-two-mile-long path ended just northeast of Mansfield.
- 23 October 1949: Almanzo Wilder dies in Mansfield, Missouri.
- 10 February 1957: Laura Ingalls Wilder dies in Mansfield, Missouri.
- 20 September 1967: Hurricane Beulah makes landfall between the mouth of the Rio Grande and Brownsville, Texas, affecting Rose Wilder Lane and her home in nearby Harlingen, Texas.
- 30 October 1968: Rose Wilder Lane dies at her home in Danbury, Connecticut.
- 24 June 2003: An F4 tornado destroys what remains of Manchester, South Dakota.

Appendix 2
Glossary of Weather and Climate Terms

Adaptation: Adjustment to environmental conditions.

Blizzard: The U.S. National Weather Service specifies sustained wind or frequent gusts of five miles per hour or greater, accompanied by falling and/or blowing snow, frequently reducing visibility to less than a quarter mile, for three hours or longer.

Chinook: A warm, dry, downslope wind descending the lee side of the Rocky Mountains as a result of broad-scale, cross-barrier flow over the mountain range.

Climate averages: Averages for climate variables like temperature and precipitation, usually for the full period of record for a location unless specified otherwise.

Climate normals: Thirty-year averages for climate variables like temperature and precipitation, updated every ten years. They provide a baseline that allows comparison of a location's current weather to the average weather that location has seen in that thirty-year baseline—whether a particular day is cooler or warmer than normal, if a particular month is wetter than normal, or if the growing season is longer than normal. At publication, the normals (baseline comparison) period is 1991–2020.

Dewpoint temperature: The temperature to which air must be cooled at constant pressure and constant water vapor content in order for saturation to occur. Used to calculate relative humidity.

Drought: A period of abnormally dry weather sufficiently long enough to cause a serious hydrological imbalance.

El Niño/Southern Oscillation (ENSO): Acronym for El Niño–Southern Oscillation, coined in the early 1980s in recognition of the intimate linkage between El Niño events and the Southern

Oscillation, which prior to the late 1960s had been viewed as two unrelated phenomena. The global ocean–atmosphere phenomenon to which this term applies is sometimes referred to as the "ENSO cycle."

El Niño: A significant increase in sea surface temperature over the eastern and central equatorial Pacific that occurs at irregular intervals, generally ranging between two and seven years.

La Niña: The most common of several names given to a significant decrease in sea surface temperature ("cold events") in the central and eastern equatorial Pacific.

Flash flood: Flooding caused by rapidly rising water level in streams, creeks, rivers, or other waterways, normally dry streambeds, or in urban areas, usually as a result of intense rainfall over a relatively small area or for moderate to intense rainfall over highly saturated or impervious land surfaces, and generally occurring within minutes to several hours of the rainfall event.

Flood: The overflowing of the normal confines of a stream or other body of water, or the accumulation of water over areas that are not normally submerged.

Forecast: An assessment of the future state of the atmosphere with respect to precipitation, clouds, winds, temperature, and moisture (dewpoint temperature or humidity).

Fujita and Enhanced Fujita scales: Relates tornado intensity indirectly to structural and/or vegetative damage.

Growing degree days: The number of degrees that the average temperature is above a baseline value. Every degree that the average temperature is above the baseline value becomes a growing degree day. Agricultural related interests use growing degree days to determine planting times.

Heat index: The heat index (HI) or "apparent temperature" is an approximation of how hot it "feels" for a given combination of air temperature and relative humidity (RH). Generally, higher RH values at the same temperature feel warmer or more stressful because of less evaporative cooling when people perspire.

Hurricane: A tropical cyclone with one-minute average surface winds in excess of sixty-four knots in the Western Hemisphere.

Hypothesis: A tentative assumption made in order to draw out and test its logical or empirical consequences.

Ice jam: An accumulation of broken river ice caught in a narrow channel.

Mitigation: An action to make something less severe.

North Atlantic Oscillation (NAO): A large-scale fluctuation in atmospheric pressure between the subtropical high-pressure system located near the Azores in the Atlantic Ocean and the subpolar low pressure system near Iceland, which is quantified in the NAO Index. The surface pressure drives surface winds and wintertime storms from west to east across the North Atlantic, affecting climate from New England to western Europe, as far eastward as central Siberia and the eastern Mediterranean and southward to West Africa.

Outlook: Issued to indicate that a hazardous weather or hydrologic event may develop. It is intended to provide information to those who need considerable lead time to prepare for the event.

Pluvial: Pertaining to rain, or more broadly, to precipitation, particularly to an abundant amount thereof.

Pyrocumulus: A cumulus cloud formed by a rising thermal from a fire, or enhanced by buoyant plume emissions from an industrial combustion process.

Red Flag Warning: A term used by fire-weather forecasters to call attention to limited weather conditions of particular importance that may result in extreme burning conditions. It is issued when it is an on-going event or when the fire weather forecaster has a high degree of confidence that Red Flag criteria will occur within twenty-four hours of issuance. Red Flag criteria occurs whenever a geographical area has been in a dry spell for a week or two, or for a shorter period, if before spring green-up or after fall color, and the National Fire Danger Rating System (NFDRS) is high to extreme and the following forecast weather parameters are forecasted to be met:

1) a sustained wind average 15 mph or greater
2) relative humidity less than or equal to 25 percent
3) a temperature of greater than 75°F.

In some states, dry lightning and unstable air are criteria. A Fire Weather Watch may be issued prior to the Red Flag Warning.

Relative humidity: Describes the amount of water vapor in the air relative to how much water vapor is in the air when it is saturated at a given pressure and temperature.

Saffir-Simpson scale: A classification scheme for hurricane intensity based on the maximum surface wind speed and the type and extent of damage done by the storm.

Severe thunderstorm: A thunderstorm that produces winds of at least fifty-eight mph and/or hail at least one inch in diameter.

Supercell: An often-dangerous thunderstorm that contains a rotating updraft, which persists for a period of time (often much longer than ten to twenty minutes).

Sustainability: Of, relating to, or being a method of harvesting or using a resource so that the resource is not depleted or permanently damaged.

Theory: A plausible or scientifically acceptable general principle or body of principles offered to explain phenomena.

Tornado: A rapidly rotating column of air extending vertically from the ground to the base of a cumuliform cloud, often with near-surface circulating debris/dust when over land or spray when over water. Although its presence is not required, a funnel cloud is often visible and may partly or fully extend from the cloud base to the ground.

Traditional Ecological Knowledge (TEK): The ongoing accumulation of knowledge, practice, and belief about relationships between living beings in a specific ecosystem that is acquired by Indigenous people over hundreds or thousands of years through direct contact with the environment, handed down through generations, and used for life-sustaining ways of existing in a particular locality.

Watch: Issued when the risk of a hazardous weather or hydrologic event has increased significantly but its occurrence, location, and/or timing is still uncertain.

Warning: Issued when a hazardous weather or hydrologic event is occurring, is imminent, or has a very high probability of occurring.

Wildfire: Any free-burning uncontainable wildland fire not prescribed for the area, which consumes the natural fuels and spreads in response to its environment.

Wind chill: The portion of the cooling of a human body caused by air motion.

Questions for Reflection and Discussion

1. What was the “big” weather event of your life? What do you remember most about it?
2. Which of Wilder’s big weather events would you want to experience alongside her, in her era?
3. What type of weather or climate hazard scares you the most? What could you do to make yourself feel safer?
4. How has the weather, climate, and environment where you live or in a place you cherish changed through the years
5. If there is one question you could get answered about weather or climate, what would it be?
6. What are some ways you could pay closer attention to the weather, climate, and environment around you, like Laura Ingalls Wilder did?
7. What are some actions you can take to help yourself, your family, and your community be better prepared for weather and climate hazards?
8. Knowing how accurately Wilder recorded the details of the weather, climate, and environment, what do you think or feel when she was not accurate or altered the story away from the reality?
9. If you could choose a place to live based only on the weather and climate, where would it be? What do you think that place will look like twenty-five years from now? Would it be different enough that you would want to move to remain in the weather and climate you chose?
10. Thinking back to Wilder’s frugality and desire to live off the land, and maybe that of your grandparents or other ancestors, what are some ways you could live more sustainably and economically?

Acknowledgments

This book is the product of over a decade and a half of research, reflection, presentations, and most of all, simple curiosity. Dozens of people deserve recognition for their help along the way, and if I have omitted anyone, please know that it is an error of my memory and documentation and not a reflection of their contribution.

The organizers of the original LauraPalooza 2010 conference, hosted by the Laura Ingalls Wilder Legacy and Research Association, took a chance on letting this weather geek come talk to their Laura Ingalls Wilder conference. Since then, hundreds of attendees have helped shape *Wilder Weather* with their questions and interest at LauraPalooza conferences in 2012, 2015, 2017, 2019, and 2022. Their curiosity is how I knew that I wasn't the only person who thought it was interesting to look at weather of the past through Laura's lens and to see what it can tell us about weather now and in the future. I'm grateful to each attendee and each organizer of all the LauraPalooza conferences. It is because of them and for them that this book even exists.

As a scientist, I've learned that when I am not an expert in something, I should find the ones who are and trust their expertise. I also know as a teacher myself how valuable it is to learn from other teachers and coaches. Caroline Malloy took a chance on me as a book coaching developmental project. I'm afraid I made her work hard for it! Jennie Nash is the best random-drawing prize I have ever won, and I am so grateful for her help to turn the screws just a little tighter to help the stories shine. Jennifer Louden helped me embrace my authentic voice and trust myself to shine a light on climate change.

Speaking of Jennifer Louden, this book would only be a pile of thoughts in my head without the safe, trusting space she created for writers at her retreats in Taos, New Mexico, in 2021–2024. I'm grateful for the wisdom of all attendees, especially my small-group members who helped me share, trust, and find my voice: Sarah Flick, Annie Graf, Mary Montanye, Jennifer Harvey, Helen Morales, Liz Norell, and Gail Marten. Thank you for helping me find, hear, and trust my voice.

A collection of Laura Ingalls Wilder experts have helped me keep my facts straight and supported this work. Pamela Smith Hill told me at LauraPalooza in 2010 that I should write a book about this, and I was floored that such a well-regarded biographer would find interest in this corner of Laura's world. Supportive words from writer to writer mean everything! Sandra Hume provided helpful feedback as a beta reader and has always been willing to converse about all things Ingalls and Wilder through the years. Nancy McCabe reviewed early drafts of my book query letters, helping me shape how I proposed the book. Bill Anderson is a font of Ingalls and Wilder knowledge and is always so eager and friendly to share it. The late John Miller inspired valuable lessons about place and time. Caroline Fraser's brilliant book *Prairie Fires* and her presentations collected valuable information about the history around Wilder's books, and I am especially grateful for her words in the Foreword. Amy Lauters, Laura Keyes, Sarah Manley, Eddie Higgins, Melanie Stringer, and the late Laura McLemore all provided tidbits of Wilder knowledge through the years, which are woven into this book. Cindy Wilson shares my curiosity about weather, Wilder, and railroads, and I'm eternally grateful for her generosity sharing data and research insights. If you like this book, pick up her books, too. "Uncle" Jim Hicks delivers master classes in science communication at each LauraPalooza, and I am grateful for our collaborations and conversations through the years.

Weather and climate experts, too, have supported this work in many ways. Martha Durr guided the early stages of this research as my Ph.D. advisor, as well as being a trusted colleague and friend. Scott Doering shares a passion for historical weather and supplied several historical tidbits. Dennis Todey at the USDA Midwest Climate Hub provided South Dakota climate insights before he moved on to cornier pastures and is now my go-to expert for all things agriculture weather and climate. Nancy Westcott at the Midwestern

Regional Climate Center fed me FORTS data every time I asked. Gilbert Compo at NOAA's Physical Sciences Laboratory patiently walked through detailed analysis of the 20th Century version 3 Reanalysis output, compared to 20th Century version 2 Reanalysis and to observations, to explore differences among the datasets, the imperfections of reanalysis, and the helpful information they contain and the tools to access them. Steve Hilberg, retired from the Midwestern Regional Climate Center, was and remains a sage in winter weather, in addition to being my AWSSI co-conspirator. Keith Berger from the National Weather Service office in Gaylord, Michigan, took me to several COOP stations and patiently explained the details of their instrument packages and histories. Ray Wolf provided an overarching review of the climate and weather in the book. Any errors or omissions in the weather and climate are my errors alone.

I'd like to give a special nod to Daryl Obermeier, a Nebraska COOP observer and farmer. While not exact to the details of his particular farm's location and climate, the farmer-observer in this book is inspired by and patterned after Mr. Obermeier's years of service as a COOP observer and a leader among his peers in climate understanding, adaptation, and mitigation.

This book leaned on the archives from the Rose Wilder Lane collection at the Herbert Hoover Presidential Library in West Branch, Iowa, as well as the Chicago & North Western archives, with information provided via email. Folks at the Michigan State Library in East Lansing made *The Long Winter* first-draft manuscript available when it was going on temporary display, even though I subsequently lost all of the images and notes I carefully collected there and had to re-access it through its home in the Burton Historical Collection at the Detroit Public Library (make sure everything is on the cloud before you wipe an iPad clean, folks!).

The South Dakota Historical Society Press has long been a valuable partner in communicating the many facets of research into Laura Ingalls Wilder's life, and I am still honored that they chose to publish my work. Thank you to press director and editor-in-chief Dedra McDonald Birzer for accepting my proposal and to her,

SDHS Press editors, and anonymous readers for their expertise and careful attention to detail, with revisions that greatly improved the manuscript.

Finally, and perhaps most of all, my family deserves an award for their support. Josh has watched me chase Wilder homesites, research documents, attend LauraPaloozas, and hone writing chops for years and years, and he has steadfastly supported every step in the journey. He's the Almanzo to my Laura, the quiet and steady rock anchoring my wanderings, emotions, surges, and lulls as I've been learning to write. Jaxon has never known life without a mom who has been working on this project or writing this book, and he even took on a big school project about Laura Ingalls Wilder just so he could learn more about this thing that has his mom's interest so captivated. He also has his eyes on weather radar as much as me (or more) these days as his passion for weather grows. Their love is my everything. My mom, Susan, has stoked the fires of all of my passion projects through the years and let my curiosity and adventures run their paths. My sister, Dawn, walked the path of weather curiosity with me when we were kids and remains a supporter of the sciences, arts, and their intersections. The curiosity that sparked this book would not exist without their fuel to my fire from my youngest years to now.

Barb Boustead
21 May 2025

Index